THE PODCASTING BLUEPRINT: UNVEILING YOUR VOICE TO THE WORLD

YOUR ULTIMATE GUIDE TO CRAFTING, GROWING, AND PROFITING FROM A CAPTIVATING PODCAST

L.D. KNOWINGS

CONTENTS

INTRODUCTION

We all have something to say. We all have something important to tell the world. Unfortunately, the problem lies in getting our message out there. Amid the vast ocean of information, broadcasts, influencers, and media, getting our message out there can be like a lone voice crying out in the desert.

If you're one of those eager to share their valuable experience, knowledge, and insights with the world, you have come to the right place. We will embark upon a journey that will help you get your message out there.

But first, let's talk about something crucial. There are ways to get your message out there. And then, there were WAYS to get your message out there. You can write a book or start a blog. You can create a YouTube channel or try your hand at Instagram influencing. While these communication formats promote success to varying degrees, they have inherent limitations.

There is one format that appeals to virtually everyone out there: podcasting.

That's right. Podcasting is the only format that seemingly appeals to everyone on the planet. You see, podcasts are a highly digestible content format. They allow users to consume significant amounts of content in a short span.

Now, let's discuss that in a little more detail.

Podcasting offers several advantages that other formats don't. For instance, reading a book or blog generally demands that users devote their full attention to it. While there's nothing wrong with that, you'll need to compete with a zillion other things vying for people's attention.

What about video content?

Video content can be a highly effective way of communicating with others. But like text-based media, your content must compete with millions of other content. Thus, making headway, especially early on, can be demoralizing.

That is why podcasting has become the go-to choice for folks looking to get their message out without competing against other content formats.

Now, you might be thinking, "Is podcasting worth my time?" This question is valid, especially if you've tried it before and got nowhere. You pour a lot of time and effort into building quality content that doesn't get any traction. Nevertheless, podcasting is the only format worth putting your time and effort into.

Here are some stats to help you visualize just how powerful podcasting can be:

- In 2023, there are close to 465 million podcast listeners worldwide. The number is expected to rise to nearly 505 million in 2024.
- The market cap for the podcasting industry is roughly $24 billion.
- In general, more than five million podcasts are currently in circulation, with more than 70 million available episodes.
- Podcasts are available in virtually every language globally, with shows in more than 150 languages.
- Video-based podcasts have become more popular than ever.
- Platforms such as Spotify and Apple podcasts dominate the podcast landscape.

With these figures, it's easy to see how podcasting has become the new frontier for a new generation of influencers and media personalities. Think of podcasting giants like Joe Rogan. He would not have become the enormous media personality he was today thirty years ago. Before podcasting, Rogan would not have had the platform he has today to get his message out there. Perhaps he might have been discouraged to get his message out there if it wasn't for today's means of communication.

You can love or hate Joe Rogan, but there's one sure thing: he has capitalized on the opportunities platforms such as YouTube and Spotify offer. Bear in mind that there is no conspiracy to prop Rogan up. He merely figured out how podcasting could catapult his message to a vast audience. This situation means that virtually anyone, with the proper focus and strategy, can leverage podcasting and communication platforms to get their message out to a worldwide audience.

But why is podcasting so much better than other means of communication?

Consider this: 22% of podcast listeners consume content while driving. Think about that for a minute. Can you read a book or watch a movie while driving? You could, but your next stop will surely be an automobile accident.

In contrast, podcasts offer a unique opportunity for audiences to consume content while they go about their daily lives. Can you think of any situation where you CANNOT listen to a podcast? You can listen to podcasts while cooking, gardening, working out, running, driving, and especially in your downtime.

Indeed, podcasts are the ideal means of communicating with anyone out there. It doesn't matter how old they are or what they do; your podcast can click with a potentially unlimited audience base worldwide.

But this cross-cutting appeal is not the only reason podcasts are so enticing. There are various other reasons why podcasts are so popular.

First of all, podcasts are particularly appealing due to their extensive variety. With a continuous stream of fresh shows emerging, something truly exists tailored for everyone. Whether your preferences lean toward politics, gaming, true crime, comedy, or anything else, podcasting offers something for everyone. Additionally, podcasting allows content creators to explore topics more deeply than text-based formats such as long-form blogging.

Also, podcasting allows content creators to build genuine relationships with their audiences. Podcasting opens the door to two-way communication in which content creators can reach their audiences with fresh, engaging content that appeals at a deeper, more personal level. While competent authors can achieve this effect with their prose, podcasting greatly facilitates building these relationships with a global audience.

But unlike writing a book, which can pose a significant entry barrier, launching a podcast is quite simple.

How so?

Taking a book from an idea to a published work can take years. This process often involves countless rejections from editors and publishers while dealing with the rigors of writing, editing, printing, and marketing a book.

That problematic process essentially disappears with podcasting. Of course, a good podcast requires time and effort to produce quality content, but launching a brand-new podcast does not involve going through all of the hoops and ladders of the publishing industry. Moreover, launching a podcast does not depend on the ever-present gatekeepers publishers utilize to keep aspiring authors in check.

With podcasting, the traditional shackles keeping authors down are non-existent. Your show's popularity depends on its merits, not what publishers think. If your show resonates with audiences, it will be a hit. Once your show finds its footing, its growth potential is virtually unlimited.

This unlimited growth potential appeals to those folks seeking to build a following. While blogs, video content, and social media can provide a solid path to building a following, their ultimate reach is limited due to their highly specialized formats.

That's where podcasts can potentially hit one out of the park. Podcasts are often the gateway to other types of services or products. As a result, a popular podcast can become a springboard to other profitable ventures. Keep in mind that the podcast is not the product itself. It is merely an introduction to other ventures content creators may have in store.

With the significant upside that podcasting offers, the challenge becomes figuring out how to get started.

That is the reason why we are here. We are having this conversation since you, like many other highly creative folks, want to get your message out to the world. While you have rightly identified podcasting as an excellent opportunity to get your message out there, you may not be sure about how to get started.

No worries. This book will take you through everything you need to know to get the ball rolling with your very own podcast. Please remember that we're not talking about pointing a camera at yourself and rambling on a topic. We're discussing building a high-quality product that accurately reflects and depicts your message. As such, the point is to focus on leveraging your talents to create the type of podcast that will generate the organic following you seek.

We will begin by understanding the nuts and bolts of podcasting. This introductory portion of our discussion focuses on laying down the foundation for what will become your brand-new endeavor. Best of all, you'll learn the tricks of the trade. Most successful podcasters don't want you to know this information is insider knowledge.

Then, we'll get into exploring and developing your podcast ideas. We'll take a broad concept and refine it into an idea you can mold into a compelling product audiences want to consume. From there, we'll investigate the hardware you need to produce your podcast. This discussion on equipment will help you assess where you currently stand and what if any, materials you'll need to acquire.

Once you're ready to roll, we'll discuss the trade secrets to recording and editing your podcast. This portion of our discussion helps you see how the pros create a pleasant listening experience that keeps listeners engaged. In other words, we focus on creating a professional-grade sound experience that appropriately reflects your content's valuable nature. Remember that your podcast is not just about recording your voice and firing off content. There must be a method to the madness. So, we also focus on how to structure your podcast episodes, helping your listeners and followers see a rationale behind the concept you have articulated as the show's overarching theme.

From there, we take a deep dive into the business side of your show. We look at your show's hosting and distribution in addition to effective marketing and promotion strategies. These elements lead to monetizing your podcast. At this point, you can turn a pet project into a profitable venture. You will ultimately have the tools to turn something you're passionate about into a potentially lucrative endeavor.

By now, you must be excited to get the ball rolling. But the big question is how to take the first step. After all, knowing how and where to get started isn't always easy.

Well, believe it or not, you have already taken that first step. By reading this book, you have made the conscious choice to embark on this fantastic journey filled with many exciting steps. This journey entails hard work and dedication. But it will become nothing short of a labor of love since your new podcast will become a true reflection of your life's passion.

What if you've already tried podcasting and haven't gotten the results you expected?

Fear not. This discussion will shed light on areas you may have overlooked before. Remember, this book is not only for those new to podcasting. This book suits anyone looking to sharpen their skills in this fulfilling field.

Don't forget: you already have everything you need to get started. Therefore, the equipment, knowledge, and expertise required to make your podcast happen is perfectly learnable. Many others have succeeded. So will you. Many successful podcasters, like Joe Rogan, started with an idea and ran with it. After finishing this book, you'll be much better equipped than Rogan was when he began.

So, if you're ready to start on the journey of a lifetime, don't hesitate. Read on to learn everything you need to get your message out. Why deprive the world of your knowledge, insights, and perspectives? There are folks out there who want to hear your message from your voice.

Don't underestimate what you have to bring to the table. You will be pleasantly surprised to find how you can make a difference in many people's lives. The time to get started is now.

Let's get on with it!

1

UNDERSTANDING PODCASTING

The beautiful thing about podcasting is it's just talking. It can be funny, or it can be terrifying. It can be sweet. It can be obnoxious. It almost has no definitive form. In that sense, it's one of the best ways to explore an idea and certainly much less limiting than trying to express the same idea in stand-up comedy. For some ideas, stand-up is best, but it's nice to have podcasts as well.

— JOE ROGAN

P odcasting is the art of unrestrained expression, where conversation takes center stage. It's a canvas without boundaries—humorous, chilling, endearing, or audacious—the absence of a fixed mold grants unparalleled freedom to delve into concepts. Unlike stand-up comedy, podcasts don't confine ideas; they let them flow, unfurl, and evolve. While stand-up has its place, podcasts offer an expansive realm for exploration. It's a harmonious blend of structured thought and unscripted discourse, enriching our understanding of ideas. With podcasting, the beauty lies in its dynamic versatility, making it an invaluable medium to unravel thoughts, fostering an intellectual and creative haven beyond confinement.

WHAT IS A PODCAST?

A podcast is a unique digital medium focused on audio storytelling and communication. Like audio capsules, podcast episodes engage, inform, and entertain listeners. The process begins with an RSS feed, a technical marvel connecting content creators and listeners, delivering fresh episodes to devices for up-to-date experiences.[1]

Unlike music playlists, podcasts mainly feature dialogue, discussions, and conversations. They offer captivating debates, interviews, and explorations of various subjects, providing new perspectives and ideas. Podcasts become personalized talk radio stations, accessible anytime, anywhere. They cover different topics, from history and science to pop culture and personal stories.[2]

Plus, accessing podcasts is simple—use devices like smartphones, tablets, or computers to subscribe to platforms like Apple Podcasts, Spotify, or Google Podcasts. With a click, you're ready to embark on a captivating audio adventure.

THE EVOLUTION AND GROWTH OF PODCASTING

The evolution of podcasting has been a remarkable journey from its inception, rooted in Apple's innovation, to its current status as a mainstream and impactful medium. From altering how we consume audio content to its profound growth during the pandemic, podcasts have transcended traditional boundaries and continue to shape how we engage with information, entertainment, and the spoken word. As podcasting's momentum continues, its future promises a tapestry of new possibilities and opportunities for creators and listeners alike.

The Apple Catalyst

Let's dive into the captivating tale of podcasts, with Apple, iTunes, and the iconic iPod at center stage. In 2005, Apple welcomed podcasting into its iTunes family, and Steve Jobs, the visionary, dubbed it the marriage of an iPod and a broadcast. While his quirky "TiVO for radio" comparison may raise eyebrows, it helps us see the bigger picture. Picture it: a digital rendezvous where your iPod and a world of fascinating chatter unite. Fast-forward to today, and this union has blossomed into a vibrant audio universe, delivering on-demand conversations and stories that shape our daily soundtrack.[3]

From Radio to On-Demand Podcasts

Before the podcast wave, there was the trusty radio—a mix of ads, unpredictable content, and beloved hosts gracing specific slots. But imagine this: podcasts enter the scene, letting you snag radio gems and groove to them at your rhythm, be it on your trusty computer or your portable sidekick, the iPod. Oh, and guess what? No more FOMO on episodes—just hit that subscribe button, and you're set. The best part? Like radio, it's a cost-free auditory adventure, offering a curated playlist of discussions, tales, and discoveries whenever and wherever you fancy.[4]

Rise of Mobile Podcast Consumption

Initially underestimated, podcasting challenged big media's dominance and transformed entertainment. Accessing on-demand content was difficult until podcasts emerged, captivating minds with their transformative audio engagement. Starting with 11% in 2009, the US podcast audience grew substantially by 2014 due to emerging technologies. This format's impact set the stage for podcasting's rapid rise, enabling easy exploration, learning, and entertainment at our fingertips.[5]

The Podcasting Renaissance during the Pandemic

During the pandemic, 60% of podcast consumption occurred at home. In 2020, 32% of Americans aged 12 and above listened to podcasts monthly, a cheerful increase from 29% the previous year, as per the Infinite Dial survey. Forecasts predict a robust surge in podcast passion, with an impressive jump to 125 million monthly listeners by 2022, up from around 100 million in 2020. This growth not only attracts more ears but also advertising dollars, transforming the podcasting landscape into a dynamic and lucrative industry.[6]

IS PODCASTING FOR YOU?

Launching a podcast offers an exciting dive into modern audio adventures akin to radio sensations. Podcasts seamlessly integrate into daily life, providing engaging learning and entertainment during routine moments. Suppose you're keen on sharing expertise, insights, or stories and connecting with a diverse audience. In that case, podcasting is a versatile platform to find your voice and turn passion into resonating audio content. Aspiring storytellers, explorers, and experts can engage in ongoing conversations with curious listeners. Embrace the learning and creation journey, for podcasting's allure lies beyond the first script, holding the potential to share something extraordinary and your unique story with the world.

Now, as you ponder whether you're truly made for podcasting, the following questions will help you figure out if podcasting is right for you:[7]

Do you have a specific aim?

Having a clear purpose for your podcast is critical. Are you looking to educate, entertain, inspire, or provoke thoughtful conversations? Knowing your aim will guide your content creation and help you connect with your target audience.

Are you a tech-savvy person?

While you don't need to be a tech genius, having some comfort with recording software, editing tools, and online platforms will make your podcasting journey smoother. Fear not, though—tech skills can be learned and improved over time!

Are you willing to spend on equipment?

Investing in quality recording equipment, such as a microphone and headphones, can significantly enhance your podcast's audio quality. It shows your dedication and ensures your listeners have a pleasant experience.

Do you have the ability to research and have an interesting perspective on a topic?

Research is the foundation of engaging content. If you enjoy diving deep into subjects, uncovering unique angles, and presenting insightful perspectives, you're on your way to creating captivating episodes.

Do you have storytelling skills?

Podcasts thrive on storytelling. Whether narrating personal experiences, crafting fictional tales, or weaving together facts into a compelling narrative, strong storytelling skills can keep your listeners hooked.

Do you like interacting with people?

Podcasting is a conversation, even if it's a one-sided one. You're on the right track if you relish connecting with listeners, interviewing guests, and fostering a community.

Do you like to share things/ideas you're passionate about and still be open to criticism?

Passion is the driving force behind successful podcasts. Being enthusiastic about your subject matter keeps you motivated. And remember, being open to constructive criticism helps you improve and refine your content over time.

In a nutshell, these questions are your compass, helping you navigate the podcasting landscape. They shed light on your strengths and areas for growth. So, if you're nodding along, feeling excitement, and ready to embrace the challenge of creating engaging audio content, podcasting might be your perfect match!

Let's now consider the flipside of this equation. The following questions and thoughts will help you figure out if podcasting isn't meant to be your cup of tea:

Are you looking to get rich overnight without putting in the work?

Podcasting requires dedication and effort. If you're seeking quick financial gains without a commitment to producing valuable content consistently, podcasting might not align with your expectations.

Are you unsure who your audience is and what they want to hear?

Knowing your audience is crucial. If you're uncertain who you're speaking to and what they want, your content might miss the mark, leaving you and your listeners unsatisfied.

Are you lacking anything unique when it comes to your show's focus?

Podcasts thrive on uniqueness. If you can't identify what sets your show apart, it may get lost in the sea of content, struggling to capture listeners' attention.

Do you think that starting a podcast is easy?

While podcasting is immensely rewarding, it's not a cakewalk. It demands planning, recording, editing, and promotion. If you believe it's a breeze, you might be unprepared for the effort required.

Do you have trouble being consistent?

Consistency is vital in podcasting. If you can't commit to a regular schedule, your audience may lose interest, and your podcast's growth might stagnate.

Are you reluctant to promote your show?

Promotion is essential for building an audience. If you're unwilling to invest time and effort in spreading the word about your podcast, it might remain hidden from potential listeners.

Are you unwilling to invest in equipment?

Good audio quality is crucial. If you're unwilling to invest in decent recording equipment, your podcast might suffer from poor sound quality, impacting the listener experience.

Do you lack self-discipline and get easily distracted?

Podcasting requires focus and self-discipline. Maintaining a podcast schedule may prove challenging if you're prone to distractions or struggle with staying on track.

Are you halfhearted about podcasting, or don't listen to any?

Passion drives podcasting. If you're not genuinely excited about the medium or don't engage with podcasts yourself, it might not be easy to sustain your enthusiasm over time.

Are you hoping to sell one product or service?

While podcasts can complement business goals, solely using them as a sales pitch might not resonate with listeners. Authenticity and value should drive your content.

Do you lack enough time?

Podcasting demands time for research, recording, editing, and promotion. If you're already stretched thin, allocating the necessary time for a podcast could be challenging.

These questions help assess your podcasting readiness. If you agree, reconsider goals or address issues before starting. Podcasting is fulfilling but requires realistic expectations and effort.

THE BENEFITS OF PODCASTING

Podcasting is your ticket to becoming an influential voice, a valuable resource, and a source of inspiration. It's a thriving, versatile platform that allows you to share your knowledge and fosters growth, connections, and opportunities.[8] So, whether you're seeking to position yourself as an industry expert, explore new revenue streams, or connect with a curious audience, podcasting offers a captivating avenue to embark upon.

So, let's consider the incredible benefits you can expect from the beautiful world of podcasting:[9]

Positioning as an Expert

Podcasts are your platform to shine. Sharing your expertise, insights, and passion can establish you as a trusted authority in your field. By consistently delivering valuable content, you build credibility and trust among your audience, helping you stand out in a crowded digital landscape.

Alternative Revenue Stream

Podcasting isn't just about sharing knowledge—it can also be financially rewarding. You can create an alternative revenue stream through sponsorships, ads, or even premium content while doing what you love.

Less Competition, More Impact

Compared to other media forms, the podcasting landscape offers a unique advantage—less saturation. With fewer competitors, your voice can resonate more effectively, allowing you to reach a dedicated audience hungry for your insights.

Growing Platform

The podcasting platform is far from stagnant. It's a dynamic landscape that's still on the rise. As the medium continues gaining traction, you can ride the wave of growth and make your mark.

Industry Influence and Thought Leadership

Podcasts are your microphone to the world. By sharing your perspective, conducting insightful interviews, and engaging in meaningful conversations, you can position yourself as an industry influencer or thought leader, impacting others and contributing to meaningful discussions.

Expanded Search Potential

Podcasts expand your online footprint. As you create relevant and engaging content, your podcast episodes become searchable, attracting new listeners and potential clients seeking your expertise.

Redirect Traffic Strategically

Strategically utilizing your podcast can lead traffic to your essential web pages. By mentioning your website, products, or services during episodes, you guide your audience to valuable destinations, fostering engagement and potential conversions.

Connecting and Interviewing

Podcasts are a gateway to connection. Through interviews, you can engage with thought leaders, experts, and intriguing personalities, forging valuable relationships and broadening your network in ways you might not have imagined.

If you're looking for a meaningful endeavor and make a little extra income, podcasting can catapult you to an entirely new level. While getting your show off the ground may take some legwork, the fact is that it is worth it.

OVERCOMING IMPOSTER SYNDROME AS A PODCASTER

New podcasters often struggle with Imposter Syndrome, feeling inadequate and questioning their value. Impostor Syndrome is a psychological phenomenon where individuals doubt their competence and feel fraudulent despite external validation.[10] This situation can lead to self-criticism, anxiety, and a reluctance to acknowledge skills. "Faking it till you make it" won't suffice in podcasting; genuine conviction is essential for success. Believing in oneself becomes infectious to the audience, making a lasting impact beyond superficial appearances. Overcoming imposter syndrome is crucial for podcasters to create authentic, enduring content that resonates with listeners.

There are five Imposter Syndrome types to consider:[11]

- **The Perfectionist.** An unrelenting pursuit of flawlessness drives the perfectionist imposter. They set impossibly high standards for themselves and often equate their self-worth with their achievements. Despite external validation, they believe any mistake or imperfection undermines their credibility, intensifying their self-doubt.
- **The Superhero.** The superhero imposter feels compelled to shoulder an excessive workload and take on multiple roles. They fear that relying on others or delegating tasks will reveal their perceived incompetence. Consequently, they may burn out as they push themselves to their limits, unable to acknowledge their limitations or seek support.
- **The Natural Genius.** Individuals with the natural genius imposter syndrome have a knack for grasping concepts quickly. However, they become anxious when faced with challenges that require effort or struggle. They fear that seeking help or needing time to master something will expose their lack of innate brilliance, leading to feelings of inadequacy.
- **The Soloist.** The soloist imposter feels compelled to accomplish tasks independently, avoiding assistance or collaboration. They view asking for help as a sign of weakness and believe their

success should be solely attributed to their efforts. This mindset can hinder growth and create unnecessary stress.

- **The Expert.** The expert imposter constantly seeks to amass knowledge and skills, feeling fraudulent if they perceive gaps in their expertise. They fear being exposed as a fraud if they don't know something, even though knowing everything is unrealistic. This perpetual quest for mastery can lead to burnout and self-doubt.

COMMON CAUSES OF IMPOSTER SYNDROME

Let's take a deep dive into what can trigger these feelings associated with Imposter Syndrome:[12]

No Entry Barrier in Podcasting

The accessibility of podcasting can inadvertently contribute to imposter syndrome. With relatively low entry barriers, anyone can start a podcast, leading some to question their legitimacy among a sea of creators. The absence of formal gatekeeping can trigger self-doubt as podcasters navigate their perceived "worthiness" in a diverse field.

Requires Multiple Technical Skills

Podcasting demands various technical skills, from recording and editing to publishing and promoting. For those unfamiliar with these facets, the learning curve can be steep. This gap in technical expertise may foster imposter syndrome, causing individuals to question their suitability and competence in managing these diverse requirements.

It Takes Time to Gain Momentum

Podcasting is a long-game endeavor requiring patience and persistence. During the initial stages, when audience growth and engagement may be gradual, podcasters might question their impact or worthiness. This impatience can fuel feelings of inadequacy, even though slow initial progress is a standard part of the podcasting journey.

Disappointed with Statistics

Podcasters often monitor statistics to gauge their success. Low download numbers or modest listener engagement can trigger self-doubt, causing podcasters to question their content's appeal or ability to create engaging episodes.

It's a Side Hustle

Many podcasters juggle podcasting alongside other commitments. This balancing act can lead to imposter syndrome, as individuals might feel that their podcast isn't a "serious" endeavor compared to their primary job. This perception may undermine their confidence and hinder their ability to embrace their podcasting role.

It Is Not Related to Their Work Experience/Education

Podcasting can provide a platform for sharing personal passions, hobbies, or interests. However, suppose a podcaster's content doesn't directly align with their professional background or education. In that case, they may experience imposter syndrome, feeling disconnected from their podcast's subject matter and doubting their legitimacy as an authority.

Remember, you're not alone. There's no need to feel ashamed or embarrassed by experiencing these feelings. Anyone who's ever been in podcasting will tell you they have gone through these emotions at some point in their careers.

TIPS ON HOW TO OVERCOME IMPOSTER SYNDROME

Overcoming Imposter Syndrome is much easier than you think. You don't need to pay expensive therapists or climb the tallest mountain to find the answer. The following tips will help focus your thoughts on what truly matters: getting your message out there.[13]

Recognize Your Feelings

Start by acknowledging and accepting your imposter syndrome feelings. Understand that it's a shared experience among many podcasters and creators. Normalizing these emotions helps you take the first step toward managing them.

Accept That You Cannot Be a Master of Everything.

Recognize that no one is an expert in everything. Embrace that podcasting encompasses various skills, and it's okay not to excel in every aspect. Instead of striving for perfection, focus on your strengths and seek help or guidance for challenging areas.

Share Your Thoughts

Don't be afraid to discuss your feelings with fellow podcasters or supportive friends. Sharing your imposter syndrome experiences can create a sense of camaraderie and remind you that you're not alone in these struggles.

Don't Compare

Avoid the comparison trap. Every podcaster's journey is unique, and their successes or challenges don't dictate your worth or potential. Focus on your progress, growth, and the value you bring to your audience.

Consider the Circumstances

Acknowledge that Imposter Syndrome often emerges in moments of growth or change. Remember that feeling out of your comfort zone is a sign of progress. Embrace these challenges as opportunities for personal and creative development.

Don't Be Scared to Fail

Failure is a natural part of any creative endeavor, including podcasting. Embrace mistakes and setbacks as valuable learning experiences that contribute to your growth. Remember, even seasoned podcasters have faced hurdles along their journey.

Enjoy the Process

Shift your focus from external validation to intrinsic enjoyment. Find pleasure in creating content, connecting with your audience, and exploring your podcasting passions. Letting go of the need for constant approval can alleviate the pressure of imposter syndrome.

BRINGING IT ALL IN

Podcasting is not just about sharing your voice; it's a journey that can lead to exciting opportunities. Podcasting can be a lucrative avenue, propelling you into the spotlight as a respected thought leader in your field. But here's the scoop—you've got to roll up your sleeves and be patient.

Podcasting is like planting a seed. It takes time to nurture and grow. Consistency, authenticity, and a sprinkle of patience are your secret ingredients. Commit to crafting compelling content, engaging with your audience, and refining your unique perspective.

As you weave your narrative, you'll watch your influence blossom. Your podcast becomes a stage for insightful discussions and connections, elevating your expertise. So, if you're up for the challenge, ready to embrace the journey, and keen on making a lasting impact, it's time to embark on the exciting adventure of developing your podcast's concept. The world is waiting to hear your voice!

2

DEVELOPING A COMPELLING PODCAST CONCEPT

I'd tell anyone who wants to be a podcaster the same thing others have recommended – find a niche you're passionate about and willing to be immersed in.

— SANDRA SEALY, CHRONICLES OF A
SEAWOMAN: A COLLECTION OF POEMS

In podcast creation, an indispensable piece of advice resonates: "Find a niche you're passionate about and willing to be immersed in." This guiding principle, echoed by seasoned podcasters, underpins the essence of the forthcoming chapter. As we develop a podcasting concept, we embark on a journey transcending mere audio content. It's a realm where passion and dedication fuse to create resonant narratives, where ideas evolve into immersive experiences for listeners.

In this chapter, we delve into the intricate process of channeling your enthusiasm into a distinct podcasting concept. We'll explore how to navigate the delicate balance between your interests and audience appeal, crafting content that speaks to your heart and captivates your listeners' ears. As we unravel the layers of niche selection, content planning, and audience engagement, you'll gain insights into transforming your passion into an authentic and compelling podcast that stands out in the bustling auditory landscape.

IDENTIFYING YOUR NICHE AND EXPERTISE

A podcast niche is a focused and tightly defined subject matter that forms the core theme of your podcast episodes. It is a compass that guides creators and listeners through the content landscape. Think of it as the unique angle or perspective that sets your podcast apart in a sea of audio offerings.[1]

Imagine you're passionate about food, and you want to create a podcast. Rather than broadly covering all aspects of food, you might choose a niche like "Plant-Based Cooking for Beginners." This niche narrows the topic to a specific audience: individuals new to plant-based cooking. By honing in on this niche, you can delve deep into the nuances, challenges, and exciting discoveries within this culinary journey.

Choosing a well-defined podcast niche is essential because it communicates a clear identity to potential listeners. When people see your podcast's title and description, they should immediately grasp the essence of what your show offers. A niche prevents ambiguity – a commitment to a particular subject establishes your expertise and builds a dedicated audience.

On the other hand, if your podcast's topic is too broad, potential listeners might not grasp its focus. For instance, a podcast, "Exploring Food," might cover everything from cooking to dining out to food history. Such a vague approach risks losing potential listeners looking for specific content. With a well-chosen niche, you attract a particular audience genuinely interested in your content, fostering a strong connection between you and your listeners.

On the whole, a podcast niche is a strategic decision that shapes your podcast's identity and appeals to a particular group of enthusiasts. It helps you craft content with depth, authenticity, and relevance, ensuring that your podcast stands out amidst the vast auditory landscape.

Tips to Choose a Podcast Niche

Choosing a podcast niche can seem daunting when first starting. After all, narrowing your focus can seem like finding a needle in a haystack. But fear not. The following pointers will help you find your footing.[2]

- **Start with a General List.** Begin by brainstorming a general list of topics that interest you. These could be anything from hobbies and interests to professional fields or social issues. The goal is to create an initial pool of ideas to work from.
- **Think about Subjects You Love.** From the general list, narrow down the topics that genuinely resonate with you. Passion is a driving force in podcasting, and hosting a show about something you love will keep you motivated and engaged in the long run.
- **Consider What Makes You Different.** Reflect on your unique perspective or angle on a particular topic. Identify aspects that set you apart, whether it's a distinct viewpoint, personal experiences, or a fresh approach that hasn't been explored extensively.
- **Consider Your Strengths and Expertise.** Think about your areas of expertise and strengths. Choosing a niche that aligns with your knowledge and skills allows you to deliver authoritative content and connect with your audience on a deeper level.
- **Think about What's Relevant in the World.** Consider current trends, societal changes, or emerging technologies capturing people's attention. A podcast that addresses timely and relevant issues can attract a larger audience.

- **Consider the Competition.** Research existing podcasts within your potential niches. Evaluate their content, format, audience engagement, and overall success. While some competition can be healthy, too much saturation might make it challenging to stand out.
- **Drill Down into Subcategories.** Once you've narrowed your general topics, drill down further into subcategories. For instance, if you're interested in health, you could focus on mental health, fitness, nutrition, or specific health conditions.
- **Compare Ideas.** With a list of potential niches, compare them based on your passion, uniqueness, expertise, relevance, and competition. Consider how each idea aligns with your goals and the value it can provide to your target audience.

Finding your niche often involves exploring a main category and branching into deeper subtopics, and for many podcasters, finding that specific niche often involves diving into various subcategories in their first few episodes before landing on a niche that resonates with their core audiences. So, don't be afraid to pivot in different directions until you find that one niche that genuinely hits the mark.[3]

CHOOSING A NICHE TOPIC

At first, you may be unsure about what topics you want to focus on. Considering the vast array of potential topics, focusing on just one or two topics can be complicated. That is why we've put together the following list of the top all-time topics for you to consider which areas might be the most suitable for your interests, expertise, and experience:[4]

- **Tech.** In this niche, you'd discuss everything related to technology, from gadgets and software to industry trends and innovations. You could review the latest tech products, provide how-to guides, and discuss the impact of technology on society.
- **Health and Fitness.** Focusing on physical well-being, this niche would cover topics like exercise routines, healthy eating habits, mental health awareness, and wellness practices. You could offer tips for maintaining a balanced lifestyle and share motivational stories.
- **Personal Finance.** This niche caters to individuals seeking financial advice, budgeting tips, investment strategies, and money management skills. You could cover retirement savings, debt management, and making informed financial decisions.

- **News.** A news-oriented podcast would provide timely updates and analysis on current events, politics, and global affairs. You could offer a unique perspective on news stories and discuss their implications.
- **Sports.** In this niche, you'd dive into the world of sports, covering games, athletes, tournaments, and sports culture. You could provide commentary, analysis, and interviews related to various sports.
- **Self-Improvement.** This niche is dedicated to personal growth and development. You could discuss topics like mindfulness, time management, goal setting, and building healthy habits to help listeners enhance their lives.
- **Business.** Catering to entrepreneurs and professionals, a business-focused podcast would offer insights into business strategies, leadership, marketing, and industry trends. You could interview successful businesspeople and share valuable lessons.
- **History.** Delving into the past, this niche would explore historical events, figures, and eras. You could provide captivating narratives, analysis of historical significance, and discussions about lesser-known stories from history.
- **Facts or Trivia.** This niche is about intriguing and lesser-known facts, trivia, and fun tidbits from various topics. You could entertain listeners with fascinating information that sparks curiosity.

- **Stories.** This versatile niche involves storytelling across genres – fiction, non-fiction, true crime, and more. You could narrate captivating stories, share anecdotes, or explore fictional worlds.
- **Comedy.** Focused on humor and entertainment, a comedy podcast could include stand-up routines, comedic sketches, satirical commentary, and lighthearted discussions.[5]
- **Culture.** Exploring cultural phenomena, traditions, and societal norms, this niche would delve into diverse aspects of cultures worldwide. You could foster cross-cultural understanding and celebrate diversity.
- **Music.** You'd explore different genres, artists, and musical trends in this niche. You could showcase new releases, discuss music history, and feature interviews with musicians.
- **Animals.** Catering to animal lovers, this niche would cover topics like pet care, wildlife conservation, animal behavior, and heartwarming animal stories.
- **Religion and Spirituality.** Focused on faith, belief systems, and spiritual practices, this niche would discuss religious topics, meditation techniques, and philosophical discussions.
- **Politics.** This niche is centered around political discussions, analysis of policies, election coverage, and debates on current political issues. You could

provide balanced perspectives on various viewpoints.

Please keep in mind that these are broad topics. As a result, you will need to drill down a bit deeper to find the perfect niche that fits your interests, expertise, experience, and value proposition. Don't hesitate to ponder what you must bring to each area carefully. The more time you take to consider your offer, the more precise your niche becomes.

DEFINING PURPOSE AND TARGET AUDIENCE

Crafting a podcast's success begins with pinpointing your target audience—a vital foundation for community building. Envision these listeners as your guiding inspiration. This mutual bond thrives on a reciprocal dynamic: tailoring unique content to a dedicated audience, igniting a symbiotic momentum. Your crafted narratives resonate deeply, while their enthusiasm propels your podcast's trajectory. Before the inaugural episode, you sow the seeds of a resonant community, nurturing a shared journey that enriches creators and listeners.[6]

The journey to discovering your purpose and target audience will take you through various stages. The challenge, therefore, becomes to navigate each stage as you paint a picture of who your intended recipients will become.

Stages to Define Purpose and Target Audience

Let's explore the various stages involved in defining your show's purpose and target audience:[7]

Define Your Niche

Choosing a niche is like discovering your podcast's identity in a vast landscape. Consider what truly interests you and what you can speak about passionately and authentically. Delve into your hobbies, experiences, and expertise to discover a unique angle that sets your podcast apart. For instance, if you're a science enthusiast, you might narrow your niche to "Astrobiology: Exploring the Origins of Life Beyond Earth."

Do Your Market Research

Market research is your compass in podcast creation. Dive into platforms like Apple Podcasts, Spotify, and other podcast directories to identify popular shows within your niche. Study their content, style, episode formats, and the engagement they generate. This research helps you identify content gaps, untapped areas, and potential opportunities to offer something new and valuable to your audience.[8]

Design an Audience Persona

Creating an audience persona is akin to crafting a character for your podcast's ideal listener. Go beyond demographics and dig into their psychographics—values, challenges, aspirations, and behaviors. This persona informs content decisions, language choice, and even episode length. For example, if your podcast caters to aspiring writers, your persona might be "Alex, a 25-year-old aspiring novelist struggling with writer's block and seeking inspiration."

Be Adaptable to a Changing Audience

Podcast audiences evolve, and staying in tune with their shifting preferences is vital. Engage with your listeners through social media, email, and podcast reviews to gather feedback. Monitor download rates and episode analytics to understand which topics resonate most. If you notice a trend of interest in a particular subtopic within your niche, consider incorporating more content to maintain relevance.

Keep in mind that your show will never be cast in stone. Instead, it is a living organism that will evolve to suit its purpose and target audience. So, don't be shy to make changes, especially as you receive listener feedback. This feedback can become a valuable source of insights that allow you to fine-tune your show's purpose.

GROWING YOUR PODCAST

Perhaps the biggest challenge most podcasters face early on is growing their show's following. This situation can be particularly complicated when you lack marketing experience. The good news is that there is no magic formula to grow your podcast. In other words, increasing your show's following requires a combination of savvy, easy-to-follow marketing tactics while patiently dedicating time to grow your following. Let's explore two critical ways in which you can take your show from a seed into a blossoming endeavor:

Organic Growth[9]

Organic growth refers to increasing your followers in real terms. This growth type means that your show adds faithful followers to its base. We're not talking about dummy accounts with fake profiles. These are living and breathing individuals who listen to your show. As a result, organic growth creates a solid foundation for your show moving forward.

Please keep in mind that organic growth can be slow. That is why many newcomers to the podcasting world become disenchanted and easily discouraged when their numbers move sluggishly. But let me tell you that organic growth is worth the effort. So, let's consider how organic growth is attainable thanks to the following tactics:

- **Consistent Quality Content.** Consistency is key. Regularly release high-quality episodes that cater to your target audience. Engaging and valuable content keeps listeners returning and can lead to word-of-mouth recommendations.
- **Optimize Your Title and Description.** Craft a compelling podcast title and description that accurately reflects your niche and appeals to your target audience. Use relevant keywords to make your podcast discoverable in search results.
- **Leverage Social Media.** Use platforms like Instagram, Twitter, Facebook, and LinkedIn to share snippets and episode announcements and engage with your audience. Participate in relevant groups and communities to connect with potential listeners.
- **Guest Appearances.** Collaborate with other podcasters or influencers in your niche by being a guest on their shows. This exposes your podcast to a new audience who already shares an interest in your topic.
- **Email Marketing.** Build an email list and send regular updates to your subscribers about new episodes, behind-the-scenes insights, and exclusive content. Email allows for direct communication with your most engaged listeners.
- **Engage with Your Audience.** Respond to comments, messages, and reviews from your listeners. Building a personal connection and

showing appreciation can foster a loyal community.

Paid Marketing[10]

If you believe paid marketing involves adding fake accounts to prop your numbers up, another thing is coming. Paid marketing is just like old-time marketing tactics. Old schoolers took out newspaper ads or paid for radio spots.

That's what paid marketing is all about.

Paid marketing focuses on advertising your show on various platforms to draw interest. Once you get people to tune in for the show, you can bank on them becoming followers. Here is a look at leveraging paid advertising to grow your show.

- **Social Media Advertising.** Platforms like Facebook, Instagram, and Twitter offer targeted advertising options. Create engaging ads with visuals, audiograms, or videos to reach your demographic.
- **Podcast Directories.** Consider investing in premium listings on podcast directories like Apple Podcasts, Spotify, and Stitcher. This can increase your podcast's visibility within search results.
- **Influencer Collaborations.** Partner with influencers in your niche to promote your podcast

to their followers. Influencers' endorsements can lend credibility and attract a relevant audience.

- **Google Ads.** Utilize Google Ads to reach potential listeners searching for topics related to your niche. You can target keywords that align with your podcast content.
- **Podcast Networks.** Join podcast advertising networks that match advertisers with suitable podcasts. This can provide access to paid sponsorships and cross-promotion opportunities.
- **Targeted Podcast Ads.** Some platforms allow you to place ads on other podcasts. Choose shows that have a similar audience to yours for maximum impact.

It should be noted that paid marketing can also become a brilliant way to boost organic growth. Remember, you're targeting living and breathing individuals. As a result, honest people who like your content will listen and follow. That is the best type of growth you can hope for. But keep in mind that patience is the name of the game.

BRANDING YOUR PODCAST

Your podcast is a brand. Like any brand, it must take on a life of its own. This life means creating a persona that resonates with your target audience and building a solid relationship with your audience. The most crucial element in your show's branding is its name. If you get the name right, you can get your foot in the door more effortlessly. If you hit and miss with the name, you might be doing your show a disservice. That is why we will discuss how you can create an excellent name for your podcast, ensuring you get it right from the start.[11]

Define Your Podcast's Content and Audience

Before brainstorming names, clearly understand your podcast's central theme, topics, and target audience. What is the core focus of your podcast? Who are you trying to reach? Consider the tone, style, and overall vibe of your show.

Brainstorm Keywords and Concepts

Gather a list of keywords, phrases, and concepts that relate to your podcast's content. These should encapsulate the essence of what your show is about. For example, if your podcast is about personal finance advice for millennials, keywords might include "money management," "millennial finance," "financial freedom," etc.

Combine Descriptive and Creative Elements

Your podcast's name should be both descriptive and creative. Combine one or more keywords or concepts from your brainstorming session with a clever twist. You want your title to stand out while still conveying what the show is about. For instance, a name like "Money Mavericks: Navigating Millennial Finances" for the millennial finance podcast balances description and creativity.

Check for Availability

Your podcast's name should be visually appealing and sound good when spoken aloud. It should be easy to remember, pronounce, and spell. This will make it easier for listeners to find and share your podcast with others.

Reflect Your Brand

Think about your podcast as a brand. Does the name align with the overall message and style of your show? Consider how your podcast name fits into cover art, promotional materials, and your podcast's website.

Technical Features to Consider

Now, let's delve into the technical side of what your podcast's branding entails:[12]

- **Episode Titles.** Just as your podcast name is essential, so are your episode titles. Make them engaging and informative, giving potential listeners an idea of the content they'll find in each episode.

- **SEO Consideration.** Think about search engine optimization (SEO) when naming your podcast. Include relevant keywords in your title to improve discoverability on podcast platforms and search engines.
- **Episode Numbers.** Decide if you want to include episode numbers in your titles. Numbering can help listeners keep track of episodes, especially if your podcast has a serialized format.
- **Subtitle or Tagline.** Some podcasts include a subtitle or tagline that provides additional context about the show's content. This can further clarify your podcast's focus.
- **Branding Elements.** Consider using a consistent naming convention for episode titles to maintain branding. This could involve using a specific format or style for episode titles.
- **Episode Length.** While not directly related to the podcast's name, indicating the episode length in the title can help listeners manage their time and choose episodes that fit their schedule.

These guidelines will help you get your show's name and brand perfect. Remember that finding your show's branding may take some adjusting. So, don't be shy to mix things up until you find the ideal identity for your show's content and personality.

Tips for Choosing the Right Name for Your Podcast

Now, let's look at some essential tips to consider when choosing your show's name and branding:[13]

- **Concise and Punchy Message.** Keep your podcast name short and impactful. Aim for a name that immediately conveys the essence of your content memorably.
- **Play with Words.** Get creative with language. Wordplay, puns, and clever combinations can make your podcast name stand out and create intrigue.
- **Prominent Brand Display.** Integrate your brand name into the podcast title. This ensures brand recognition and helps listeners associate your content with your brand.
- **Establish Tone of Voice.** Your podcast name should reflect the tone and style of your show. Whether formal, conversational, humorous, or serious, the name sets the tone for what listeners can expect.
- **Search Usability.** Choose a podcast name that's easy to search for. Avoid overly obscure or generic names that might get lost in search results on podcast platforms and search engines.

- **Word-of-Mouth Referrals.** A catchy and unique podcast name is more likely to be remembered and shared by word of mouth. This can significantly boost your show's visibility and audience growth.
- **Read Aloud Often.** Test how your podcast name sounds when spoken aloud. It should roll off the tongue and be easy to pronounce, making it memorable for listeners.
- **Network Feedback.** Contact your network, friends, family, and colleagues for their thoughts on your podcast name options. Fresh perspectives can provide valuable insights.
- **Consider Visuals.** Imagine how your podcast name will look in cover art, social media banners, and promotional materials. It should be visually appealing and aligned with your overall brand.
- **Feedback Loop.** As you narrow down your choices, keep refining and iterating. Share your potential podcast names with a small group of trusted individuals and gather their feedback to decide.

BRINGING IT ALL TOGETHER

Discovering your podcast's identity is paramount for a successful launch. A compelling identity resonates with your audience, drawing them in with a clear understanding of your content's value. However, branding is dynamic and should evolve. Fearlessly experiment with crafting an image that aligns with your personality, content, and audience.

Embrace change to refine your podcast's essence, ensuring it remains relevant and captivating. A strong identity captivates listeners and establishes a loyal community, setting the stage for enduring success.

3

ESSENTIAL PODCASTING EQUIPMENT AND SOFTWARE

Choose your tools carefully, but not so carefully that you get uptight or spend more time at the stationery store than at your writing table.

— NATALIE GOLDBERG

Selecting the right tools for any endeavor is vital, but a balance must be struck. The essence of productivity lies in the choice of tools and in applying those tools. As the above quote suggests, while choosing your tools thoughtfully is essential, becoming overly meticulous can hinder progress. Whether in podcasting or any creative pursuit, the focus should remain on the task rather than acquiring the perfect equipment.

It's easy to get caught up in the allure of new stationery or the latest gadgets. Still, the true magic happens at the writing table. Time spent crafting, creating, and refining shapes the outcome. Thus, choose your tools carefully, but don't let the quest for perfection overshadow the art itself. Success emerges from a harmonious blend of thoughtful selection and dedicated action.

In this chapter, we will look at the tools of the trade. This discussion will provide a comprehensive look at the equipment you need to produce a professional-quality podcast leading to the best possible listening experience.

ESSENTIAL RECORDING EQUIPMENT AND SOFTWARE

Creating a podcast has become remarkably straightforward with the essential tools that you might already possess. To embark on this exciting journey of audio content production, you'll need a computer, microphone, headphones, and recording software. These components empower you to craft captivating episodes for your listeners' enjoyment.

A quality microphone ensures that your voice is captured with clarity and depth. At the same time, headphones enable you to monitor your recordings effectively. Your computer serves as the central hub for content creation, where recording and mixing software harmonize your audio elements. Reliable internet access is crucial in reaching your audience by facilitating episode distribution.[1]

The beauty of podcasting lies in its accessibility. With the gear above, you can delve into spoken expression, sharing your ideas, stories, and expertise with a global audience. This practical setup minimizes barriers to entry and emphasizes the potential for creative engagement and meaningful connections through podcasting.

So, let's take a closer look at the materials you will need to get your show off the ground:

Computers

Modern computers are well-equipped for podcast recording. It's likely sufficient for typical podcasting needs if you've acquired a computer recently. No immediate need exists for purchasing new hardware. Utilize your current computer – if it performs adequately, that's perfect. Over time, if you find it lacking in your requirements, consider upgrading to a model with enhanced memory and processing speed. This approach ensures you're optimizing your resources and investment. Remember, there's no rush to replace your computer; focus on content creation. When your podcasting demands grow, a more robust system can be a thoughtful investment.[2]

Microphones[3]

Basic microphone equipment:

When venturing into recording, a fundamental tool you require is a microphone to capture your voice. While top-tier options offer superior quality, your choice should align with your podcasting goals. Investing in better quality yields more professional audio, enhancing listener engagement. Remember, subpar audio quality can deter potential listeners.

USB microphones, tailored for computer compatibility, are a hassle-free choice. Their plug-and-play functionality simplifies setup, making them ideal for beginners. For solo podcasting endeavors, USB microphones usually suffice. They're well-suited for capturing single-person conversations, interviews, or narrations.

Consider your objectives and budget when selecting a microphone. USB microphones grant an efficient entry point if initial quality meets your standards. As your podcast evolves, you can explore higher-quality options. Remember, a microphone's role in shaping your podcast's auditory experience is pivotal, influencing your audience's perception and engagement.

High-End Microphones:

Upgrading your microphone setup can elevate your production quality as you progress in your podcasting journey. Transitioning to a microphone with an XLR connection offers greater flexibility and control. Still, it requires an audio interface or mixer. These additions empower you to tune your recordings finely. Some microphones provide USB and XLR options, allowing for a gradual transition. Begin with the convenient USB connection and later integrate a mixer or audio interface for XLR capabilities.

Microphones fall into two categories: dynamic and condenser. Dynamic microphones excel in minimizing feedback and are resilient in non-soundproof environments. They are budget-friendly but offer a narrower dynamic range. In contrast, condenser microphones are more sensitive, boast a broader dynamic range, and come at a higher cost.

Microphone sound pickup patterns affect audio capture, including omnidirectional, bidirectional, and cardioid. A cardioid microphone is ideal in less soundproof settings, focusing solely on front-facing audio. For co-hosting scenarios, a bidirectional pattern suits best.

While the options might initially seem overwhelming, versatile microphones combining USB/XLR connections, dynamic/condenser functionalities, and various pickup patterns simplify your selection process. Tailor your choice to your specific requirements, ensuring your podcast's sound quality aligns with your evolving aspirations.

Recommended brands by experts: [4]

- ATR2100-USB
- Blue Yeti USB Mic
- Heil Sound PR 40 Dynamic Cardioid Studio Microphone
- MXL 990 Condenser Microphone
- Pyle PDMICR42SL Classic Retro Microphone
- Shure MV7

Headphones

When recording, headphones become a crucial tool for monitoring the sound as it's captured. However, not all headphones are created equal. Opt for hard-shell headphones over their soft-shell counterparts, which only sport foam exteriors. Soft-shell headphones fail to suppress sound, potentially leading to feedback issues adequately. In contrast, hard-shell headphones featuring sturdy plastic or rubber exteriors effectively trap sound, enhancing your monitoring experience.[5]

While breaking the bank on headphones is unnecessary, remember that quality corresponds to investment. Inexpensive options might provide subpar sound quality. If high-fidelity sound matters to you, allocate resources accordingly. This becomes especially relevant if your podcasting journey evolves into multitrack audio mixing. Discriminating headphones allow precise audio tweaking, ensuring your final product meets professional standards.

Consider your preferences and aspirations when selecting headphones. Your choice should harmonize with your podcasting goals, ensuring accurate sound monitoring during recording and facilitating future audio refinement if needed.

Recommended brands by experts:[6]

- Audio-Technica ATH-M30x
- Audio-Technica ATH-M50x
- Beyerdynamic DT 770 PRO
- Sennheiser Momentum 3.0
- Shure SRH440
- Sony MDR1AM2/B

Audio Interface

Constructing a podcast studio entails crucial hardware, and an integral yet compact component is the audio interface. Functioning as a vital link, the audio interface transforms analog audio signals from your microphone into digital format. This conversion enables seamless transmission to your computer, facilitating playback, mixing, and eventual upload of your podcast episodes.

Though USB microphones can be directly plugged into computers, some podcasters opt for the added enhancement of an audio interface. By doing so, they preserve the pristine quality of the initial audio capture. This strategy contrasts with relying solely on the computer's built-in sound card, which might compromise the integrity of critical audio conversion.

The audio interface bridges the gap between analog and digital realms, underpinning the creation of professional-grade podcast content. Assembling a podcasting setup with meticulous consideration of components like the audio interface ensures optimal audio fidelity and sets the stage for a captivating auditory experience for your listeners.

Recommended brands by experts:

- Behringer XENYX 1204USB
- Focusrite Scarlett 2i2 Studio 3rd Gen USB Audio Interface Bundle
- Mackie ProFX10v3
- Midiplus Smartface II Audio Interface
- MOTU 8pre USB

Mixers

Distinguishing between an audio interface and a mixer is pivotal for optimizing your podcasting setup. While both serve to enhance audio quality, they fulfill different roles. Like an audio interface, a mixer empowers you with increased control over sound levels, inputs, outputs, and audio manipulation. Many digital audio workstations (DAWs) allow mixing and mastering multiple audio channels directly on your computer.[7]

While most podcasters rely on DAWs for mixing, some opt for manual control through a mixer, driven by personal preference. Additionally, circumstances like remote call-in guests might necessitate using a mixer for real-time audio management.

Behringer, Mackie, and the Focusrite Scarlett series are notable options to explore when considering a mixer. These brands offer mixers that cater to diverse podcasting needs, ensuring seamless integration with your recording environment.

Understanding the nuances between audio interfaces and mixers empowers you to tailor your podcast setup to your preferences and requirements. Whether you choose the precise control of a mixer or the flexibility of a DAW, your choice lays the foundation for delivering top-notch audio content to your listeners.

Recommended brands by experts:

- Behringer Xenyx 802
- Rode RODECaster Pro

Optional Accessories:

Video Cameras

If you're keen on video podcasting, the choice of equipment becomes paramount. Incorporating a video element adds depth and engagement to your content. Investing in a dedicated video camera or camcorder is highly recommended to achieve the best results.

These specialized devices are optimized for capturing high-quality visuals and offer a range of features tailored to video recording. They precisely control settings like focus, exposure, and framing, ensuring your podcast's polished and professional appearance. Utilizing a dedicated video camera or camcorder elevates the overall quality of your podcast, enhancing its visual appeal and keeping your audience captivated.

While smartphones or webcams can serve in a pinch, a purpose-built video camera or camcorder guarantees superior image quality, flexibility, and control. By choosing this, you're setting the stage for an exciting, visually immersive, and ultimately perfect podcasting experience that resonates with your viewers.

Recommended brands by experts:[8]

- Kickteck Full HD 1080p
- Nikon COOLPIX B500
- Panasonic HC-V770 Camcorder

- Sony HDR-CX405/B

Lighting[9]

When it comes to video podcasting, lighting plays a pivotal yet often overlooked role in shaping the quality of your content. Effective lighting enhances the visual experience, enabling your audience to perceive emotions and body language, especially during extended interviews or discussions.

Investing time and effort in achieving outstanding lighting setups pays dividends in terms of audience engagement and perception. Subtle nuances of facial expressions, gestures, and reactions become more pronounced, adding depth and authenticity to your podcast.

Moreover, professional-looking lighting doesn't just elevate the visual aspect of your content—it's also a potent tool for attracting new viewers and listeners. A polished and well-lit video exudes credibility and seriousness, conveying a sense of professionalism that resonates with your audience. First impressions matter, and a visually appealing video is a powerful means to make a substantial impact.

Understanding the significance of lighting and making deliberate choices to optimize it can significantly enhance the effectiveness and appeal of your video podcasts.

Recommended brands by experts:[10]

- Fancierstudio Lighting Kit 2400 Watt Professional Video Lighting Kit
- GVM RGB LED Video Lighting Kit, 800D Studio Video Lights with APP Control
- Neewer 2 Packs Dimmable Bi-Color 480 LED Video Light and Stand Lighting Kit

Pop Filters

Pop filters are indispensable tools in the world of audio recording, serving to rectify issues across the audio spectrum. By mitigating both high and low-frequency anomalies, they facilitate smoother and more seamless editing, resulting in a final recording that's easy to work with and pleasing to the ears.[11]

Prominent models like the Auray PFSS-55 Pop Filter and Avantone PS-1 Pro-Shield offer more than just their core function. Beyond preventing plosive sounds, they provide a layer of protection for your microphone. Shielding the microphone from saliva and moisture safeguards against potential buildup that could degrade the audio quality or even damage the microphone over time.

Incorporating a pop filter into your recording setup not only contributes to the technical quality of your recordings but also promotes the longevity of your microphone. This small investment translates to a more efficient post-production process and enhanced audio output, making your content more engaging and professional to your listeners.

Recommended brands by experts:[12]

- Nady MPF-6
- Shure PS-6 Popper Stopper

Shock Mounts[13]

As far as audio recording goes, shock mounts emerge as necessary accessories that address various sources of unwanted noise. They play a dual role by mitigating disturbances from accidental microphone contact and countering the vibrations transmitted from moving microphone stands or arms.

Recognizing the diversity in shock mount quality is important, emphasizing the value of allocating more resources for a better product. While shock mounts might appear straightforward, their design and construction significantly impact their effectiveness and longevity.

Budget shock mounts often feature subpar materials like low-quality rubber or plastic, rendering them susceptible to becoming brittle or fracturing over time. Moreover, their universality might be compromised, resulting in loose fits for more miniature microphones or the inability to accommodate larger equipment.

Investing in a well-constructed shock mount is an investment in your audio quality and recording setup's durability. Premium shock mounts offer sturdier construction, universal compatibility, and the assurance of sustained performance over the long haul. Choosing wisely when it comes to shock mounts ensures cleaner, more professional audio recordings while safeguarding the integrity of your recording equipment.

Recommended brands by experts:

- LyxPro MKS1-B Condenser Spider Microphone Shockmount
- Rode PSM1 Shock Mount
- Rycote 44901 Invision USM
- Sabra Som SSM-1

Portable Digital Recorders[14]

A portable digital recorder is a must-have tool for podcasting on the go. While smartphones can serve this purpose, opting for a dedicated digital recorder offers distinct advantages, especially when considering power longevity and audio quality.

Portable digital recorders are designed to withstand extended recording sessions, ensuring you won't run out of power during critical moments. This reliability is essential when capturing content while traveling, where access to charging might be limited.

Furthermore, the superior recording quality provided by digital recorders makes a noticeable difference in the final product. They're optimized to capture clear, crisp audio with minimal interference, enhancing the overall listening experience for your audience.

While phones offer convenience, their primary function is not podcasting, and various factors might compromise audio recording. Investing in a portable digital recorder supports consistent performance and higher audio standards. This choice empowers you to capture high-quality content during your travels. It ensures your podcast maintains its professional edge, regardless of your location.

Recommended brands by experts:[15]

- DYW Portable Digital Voice Recorder
- TASCAM DR-05X Recorder
- ZHKUVE Rechargeable Sound Audio Recorder

Editing and Recording Software

Editing and recording software are essential in podcast production, serving distinct but interconnected purposes. Here's how they contribute to creating a polished podcast:

Recording Software

Recording software, often called a Digital Audio Workstation (DAW), captures the raw audio during the recording phase. It's the platform where you record your podcast episodes. Critical functions of recording software include:

- **Audio Input.** It interfaces with your microphone or other audio sources, allowing you to capture your voice and other audio elements.
- **Track Management.** Recording software enables you to work with multiple audio tracks. This is particularly useful for podcasts involving numerous speakers, interviews, or adding background music.
- **Real-time Monitoring.** Most recording software provides real-time audio monitoring, allowing you to hear yourself as you record, which helps maintain audio quality and consistency.
- **Basic Effects.** While the primary function is recording, many DAWs also offer basic effects like noise reduction, compression, and EQ adjustments during recording.

Editing Software[16]

Editing software comes into play after recording, enabling you to refine the recorded audio, arrange different elements, and create the final version of your podcast episode. Editing software provides several critical capabilities:

- **Audio Editing.** You can cut, copy, paste, and arrange audio segments, ensuring a smooth and coherent episode. This includes removing mistakes, long pauses, or any unwanted background noises.
- **Enhancements.** Editing software allows you to apply advanced effects and adjustments like noise reduction, equalization, compression, and audio normalization to enhance audio quality.
- **Adding Elements.** You can seamlessly insert intro/outro music, ads, interviews, or pre-recorded segments into your podcast.
- **Mixing.** Balancing audio levels, adjusting panning (stereo placement), and controlling the overall sound mix ensure that all components are audibly coherent.
- **Exporting.** After editing, the software facilitates exporting the final episode in various formats (MP3, WAV, etc.) suitable for distribution to podcast platforms.

As you can see, recording software helps capture the raw audio, while editing software refines and polishes that audio into the final podcast episode. Both are crucial for maintaining audio quality, creating engaging content, and presenting a professional product to your listeners.

Recommended recording and editing software include:

- Alitu
- Audacity
- Garageband
- Hindenburg Journalist
- Podcastle
- Ringr

Hosting Software[17]

Podcast hosting software is vital in producing and delivering your podcast to listeners. It bridges your finished podcast episodes and the platforms where listeners can access and subscribe to your content. Here's how podcast hosting software works:

- **Uploading and Storing.** Once your podcast episodes are edited and finalized, you upload them to the podcast hosting platform. The hosting software stores these files on their servers. These servers are designed to handle the distribution of your podcast to various podcast directories and apps.
- **Generating RSS Feed.** Podcast hosting software creates your podcast's RSS (Really Simple Syndication) feed. This feed is a specialized XML file containing information about your episodes, including titles, descriptions, publication dates,

and media file URLs. Podcast directories and apps use this RSS feed to fetch and display your episodes to listeners.

- **Distributing to Directories and Apps.** Podcast hosting software automatically distributes your podcast episodes to various podcast directories such as Apple Podcasts, Spotify, Google Podcasts, and more. It ensures that your episodes are available to listeners on their preferred platforms.
- **Tracking Analytics.** Hosting software provides analytics and insights into your podcast's performance. It tracks metrics like the number of downloads, listener locations, popular episodes, and audience growth over time. These analytics help you understand your audience and make informed content decisions.
- **Ensuring Reliability and Scalability.** Podcast hosting services offer reliable and scalable server infrastructure. This ensures that your podcast episodes are accessible to listeners without interruption, even during high traffic or popularity periods.
- **Managing Show Information.** Hosting software allows you to manage your podcast's metadata, including show title, description, cover art, and author information. Any changes you make are reflected in the RSS feed and podcast platforms.

On the whole, podcast hosting software acts as a central hub for your podcast episodes, creating an RSS feed for distribution, ensuring accessibility across platforms, providing analytics, and offering the reliability needed to deliver your content to your audience. Without podcast hosting, your episodes wouldn't reach listeners through their preferred podcast apps and directories.

Here is a look at the recommended hosting software:

- Buzzsprout
- Simplecast
- Transistor

All-In-One Podcast Software[18]

All-in-one podcast software is a comprehensive solution that streamlines various aspects of podcast production into a single platform. It aims to simplify the podcasting process by integrating recording, editing, hosting, distribution, analytics, and sometimes even monetization features.

This software allows you to eliminate the need for individual software suites by combining the following functions into a single app you can control from a centralized dashboard:

- Recording and editing
- Audio processing and effects
- Hosting and distribution
- Analytics and insights
- Monetization tools
- Customization and branding
- User-friendly interface

Above all, an all-in-one software solution eliminates the need to switch between different tools and services. This saves time and effort during the production and distribution process.

Recommended all-in-one software solutions include:

- Anchor
- Cast

Digital Audio Workstations (DAWs)

Digital Audio Workstations (DAWs) are essential when producing a podcast, offering a comprehensive environment for recording, editing, and arranging audio content. They are software applications designed to facilitate the creation and manipulation of digital audio. Here's what DAWs do in the podcast production process:

- Recording
- Editing

- Multitrack mixing
- Effects and processing
- Automation
- Adding music and sound effects
- Exporting
- Plugins and virtual instruments
- Mastering

As you can see, digital audio workstations are versatile software applications that empower podcast producers to create, edit, and refine audio content. They provide a comprehensive toolbox for enhancing audio quality, crafting engaging episodes, and delivering a polished podcast to your audience.

Recommended DAWs include:

- Adobe Audition
- Avid Pro Tools

HOW TO SET UP YOUR RECORDING STUDIO

Setting up a podcast recording space requires careful consideration to ensure excellent audio quality. Here's a detailed guide for each step:[19]

Choose the Right Space

- **Location.** Pick a quiet room with minimal external noise. Avoid areas near busy streets, noisy neighbors, or continuous sound sources. Ideally, opt for an interior room with fewer windows and openings.
- **Size.** Smaller rooms with less space for sound to bounce around are generally better. Large rooms can create echoes and reverb, which can be challenging to control.
- **Furniture and Layout.** Add soft furniture like couches, cushions, and curtains to help absorb sound reflections. Arrange furniture to reduce sound reflections by avoiding parallel walls and surfaces.

Soundproof Your Podcast Studio Setup

- **Acoustic Treatment.** Use acoustic panels, bass traps, and foam to control sound reflections. Place panels on the walls, corners, and ceiling to minimize reverb and echoes. This improves the overall audio quality and reduces unwanted noise in your recordings.
- **Sealing.** Check for any gaps around doors, windows, and vents. Seal these openings to prevent external noise from entering and your recordings from leaking out.

- **Flooring.** If possible, use carpets or rugs to absorb sound further. Hard surfaces can cause sound to bounce, leading to unwanted reflections.

Grab Your Recording Equipment

- **Microphone.** Choose a quality microphone that suits your recording environment and vocal characteristics. Consider dynamic microphones for less sound leakage and condenser microphones for capturing detail.
- **Headphones.** Select closed-back headphones to minimize sound leakage into the microphone. This helps you monitor your audio more accurately during recording.
- **Pop Filter.** Attach a pop filter to your microphone stand to prevent plosive sounds ("p" and "b" sounds) from distorting your recordings.
- **Microphone Stand or Boom Arm.** Stabilize your microphone to avoid handling noise. A stand or boom arm ensures consistent microphone placement.
- **Shock Mount.** If your budget allows, use a shock mount to isolate the microphone from vibrations and bumps, further improving recording quality.

Pick Your Podcast Editing Software

- **Choose a DAW.** Select a Digital Audio Workstation (DAW) that suits your needs and proficiency level. Popular options include Audacity (free and user-friendly), Adobe Audition, GarageBand (for Mac users), or Reaper.
- **Familiarize Yourself.** Learn the basics of your chosen DAW. Understand how to import audio, cut, paste, apply effects, and export your final podcast episode.

Choose a Podcast Hosting Service

- **Research Hosting Providers.** Research podcast hosting services that align with your goals. Look for platforms that offer reliability, scalability, analytics, and distribution to major podcast directories.
- **Sign Up and Upload.** Sign up for your chosen hosting service, create your podcast profile, and start uploading your episodes. The hosting service will provide an RSS feed that allows platforms like Apple Podcasts, Spotify, and Google Podcasts to access your episodes.

BRINGING IT ALL TOGETHER

Setting up everything you need is a pivotal stride toward crafting a professional-grade podcast. Dedication to creating the right recording environment is vital, demanding both time and resources. By meticulously establishing an optimal space, you lay the foundation for delivering a superior listening experience to your audience.

Your investment in setting up the perfect podcasting space reflects your commitment to quality. It ensures that your content resonates with listeners in the best possible way.

4

RECORDING AND EDITING YOUR PODCAST

Don't bore people. Don't worry too much about replicating someone else's formula. Be original with the way you podcast.

— JAMES SCHRAMKO

When it comes to podcasting, captivating your audience is paramount. As the quote wisely advises, "Don't bore people." The podcasting landscape is brimming with content, and to stand out, you must strive for uniqueness. While seeking inspiration is essential, don't be overly concerned with mimicking someone else's success. Instead, focus on crafting an original and authentic podcasting experience.

Originality is the key to leaving a lasting impact. Your voice, perspective, and ideas are your assets. Use them to create a distinctive podcast that resonates with your target audience. Embrace your individuality and explore topics that ignite your passion. Inject your personality into your episodes, and don't shy away from taking creative risks.

Remember, listeners seek fresh insights and engaging discussions. By daring to be different, you'll avoid monotony and attract a loyal following eager to tune in for your distinct take on subjects. So, as you embark on your podcasting journey, heed this advice: Be original, be daring, and above all, don't hesitate to infuse your unique essence into every episode.

With originality in mind, let's focus on recording and editing your podcast. Remember that even excellent content can be ruined by poor recording quality and ineffective editing. So, this chapter will set you up with the tools and strategies you need to create a top-quality podcast.

MICROPHONE TECHNIQUES FOR QUALITY SOUND

Proper microphone technique is paramount for achieving top-notch sound quality. It's the foundation upon which clear, professional audio is built. Maintaining the correct distance, avoiding handling noise, utilizing appropriate accessories, and understanding microphone characteristics are crucial. These techniques capture your voice precisely, free from distortion, ambient noise, and unwanted artifacts.[1]

With proper technique, your podcast, recording, or performance gains a polished, engaging quality that captivates listeners and sets the stage for a remarkable auditory experience. That is why the following considerations are crucial to ensuring you produce professional-quality audio:[2]

Choose the Right Microphone

Selecting the right microphone involves understanding the specific characteristics that match your recording environment and voice. Dynamic microphones are durable and handle high sound pressure levels, making them ideal for recording in noisy surroundings or on-location interviews. Condenser microphones, on the other hand, offer a broader frequency response and greater sensitivity, which is valuable in controlled studio settings. Consider your recording space, the type of content you produce, and the tonal qualities of your voice when making your choice.

Find the Right Distance

Placing the microphone at the appropriate distance from your mouth is crucial. Too close, and you risk distortion; too far, and your voice might sound distant and weak.

Aim for a 6 to 12 inches distance, adjusting it slightly based on the microphone's characteristics and your speaking volume. Experimentation will help you find the sweet spot that captures the full richness of your voice without overloading the microphone.

Maintain a Constant Distance

Consistency in microphone placement is vital to ensure uniform audio quality. Avoid moving in and out during recording sessions, resulting in uneven volume levels. A microphone stand, boom arm, or a hands-free setup will help you maintain a steady distance, producing more professional-sounding content.

Keep Your Hands off the Mic

Touching the microphone during recording introduces handling noise, which can distract your final recording. Employ a shock mount, a device that suspends the microphone, to isolate it from vibrations transmitted through surfaces. Additionally, practice good microphone discipline by refraining from tapping, tapping, or bumping the microphone inadvertently.

Dampen the Room Sound

Achieving a clean audio recording involves minimizing room reflections and echoes. Acoustic treatment materials like foam panels, bass traps, diffusers, and even blankets can help absorb and diffuse sound waves, reducing unwanted reverb. Experiment with microphone placement and the strategic positioning of acoustic treatments to find the best setup for your space.

Use Separate Mics for Each Person

When hosting a multi-person podcast, providing individual microphones is crucial. Sharing a single microphone can lead to uneven audio levels and make postproduction editing challenging. Using separate microphones ensures clear and distinct recordings for each participant, resulting in a more professional and polished end product.

Stand on Your Feet

Opting to stand while recording can significantly enhance your vocal performance. Standing encourages better posture and allows your diaphragm to function more effectively, leading to improved breath control and vocal projection. This technique helps you sound more energetic and engaged in your podcast episodes.

Use a Pop Filter

A pop filter is a screen placed in front of the microphone to soften plosive sounds caused by intense bursts of air hitting the microphone diaphragm during "p" and "b" sounds. This accessory prevents sharp bursts of air from distorting the recording, maintaining a consistent and pleasing audio quality.

Angle Your Mic

Positioning the microphone slightly off-axis to your mouth helps minimize direct exposure to plosive sounds and sibilance, which are common in speech. Experiment with different angles to find the optimal position that reduces these audio anomalies while preserving the natural qualities of your voice.

Always Wear Headphones

Monitoring your recording using headphones lets you identify and address issues in real time. This includes detecting background noise, distortion, or any technical glitches that might affect the quality of your recording. Wearing headphones ensures that you're capturing the cleanest audio possible.

Limit Ambient Sounds

Before recording, take steps to eliminate potential sources of ambient noise. Close windows and doors to block out external sounds, and turn off any appliances or fans that could introduce background noise. This creates a controlled acoustic environment, producing a more professional and polished final recording.

By incorporating these meticulous microphone techniques into your podcasting routine, you'll be well-equipped to produce content that resonates with your audience due to its impeccable audio quality. Remember that mastering these techniques takes time and practice, but the payoff for improved sound is worth the effort.

RECORDING TIPS AND BEST PRACTICES FOR BEGINNERS

You may encounter some difficulties as you gain experience in the podcasting world. But fear not. The following pointers will help you hit the ground running immediately.

Use the Right Equipment

Invest in a good-quality microphone, headphones, and a pop filter to capture clear and clean audio. A USB or XLR microphone with an audio interface can be excellent for beginners.

Don't Forget to Warm Up

Like an athlete warms up before a game, spend a few minutes warming up your voice before recording. This helps in achieving consistent vocal tone and clarity.

Record in a Small, Quiet Room

Choose a small room with minimal ambient noise to prevent unwanted echoes and distractions. Consider using blankets, pillows, or foam panels to dampen the room's sound.

Create a Brief Noise Profile

Record a few seconds of silence before you start speaking. This allows you to capture the ambient noise in the room, which can later be used to reduce background noise during editing.

Adopt Proper Microphone Techniques

Maintain a consistent distance from the microphone, use a pop filter to minimize plosives, and handle the microphone gently to avoid making noise.

Watch Your Volume Levels

Aim for consistent audio levels without clipping (distortion from high volumes) or being too quiet. Use a pop filter and adjust your microphone's gain appropriately.

Watch Your Breath

Be mindful of heavy breathing or sighing into the microphone. Breathe naturally and consider editing out any overly loud breaths during post-production.

Keep Your Body Still

Minimize movements that could introduce unwanted noise. Sit or stand still during recording to maintain a consistent audio quality.

Resolve Sound Issues Early

If you notice technical glitches or sound issues, address them immediately to avoid complications during editing. It's easier to fix problems in real-time than during post-production.

Record with Headphones (Your Guest Too)

Wearing headphones allows you to monitor audio quality and catch any issues while recording. Ensure your guest also wears headphones to prevent audio bleed.

Stay Quiet While Your Guests Speak

Avoid background noise or speaking while your guest is talking. This ensures clean recordings and makes editing smoother.

Leave Audio Cues for Mistakes

If you make a mistake during recording, pause and clap your hands or make a distinct noise. This creates a visible spike in the audio waveform, making it easier to identify and edit out mistakes later.

Mute When You Aren't Speaking

If you're not talking, mute your microphone to eliminate any accidental noises you might make, like coughing or shuffling papers.

Use Production Elements Sparingly

Background music, sound effects, and transitions can enhance your podcast, but use them judiciously to avoid overwhelming the listener.

Stay Hydrated[3]

Keep a glass of water handy to prevent dry mouth, which can affect your voice quality.

Record a Separate Channel for Each Person

Record each person on a separate audio channel if recording with multiple people. This allows for individual editing and ensures balanced audio.

Record Under a Blanket

Recording under a blanket or using a pillow fort can help create a makeshift sound booth by reducing echoes and external noise.

Don't Be Afraid to Take a Break

Recording for extended periods can lead to fatigue and decreased vocal quality. Take short breaks to rest your voice and regain energy.

Maximize Your Internet Bandwidth

If conducting remote interviews, ensure a stable internet connection to prevent dropouts and audio artifacts.

Trust Your Ears and Take Notes

While recording, pay attention to any issues or areas that need improvement. Trust your instincts and make notes for editing later.

Don't Forget About the Content

While technical aspects are important, the content is the heart of your podcast. Prepare, research, and engage with your subject matter passionately.

Mastering these podcast recording tips and best practices will enable you to produce high-quality, engaging content that resonates with your audience. Remember that practice, patience, and continuous improvement are vital to achieving podcasting excellence.

EDITING AND POST-PRODUCTION TECHNIQUES

Once you have successfully recorded your content, the editing and post-production process allows you to filter issues that may hinder your audience's overall listening experience. So, let's consider the aspects during the editing and post-production process.[4]

Stage 1: Podcast Editing

Podcast editing is a pivotal stage that transforms raw audio recordings into polished, engaging episodes. Let's delve into the intricacies of this process:

- **Define the Length of Your Podcast Episodes.** Determining the optimal length for your podcast episodes is essential. Consider your content, audience preferences, and the depth of your subject matter. Shorter episodes, around 15-30 minutes, are suitable for quick insights, while longer ones of 45-60 minutes allow for more in-depth exploration. Consistency in episode length helps manage audience expectations and maintain engagement.
- **Create a Compelling Story Through Podcast Editing.** Editing is where your podcast takes shape. Craft a compelling narrative by arranging your content in a logical sequence. Trim excessive tangents and repetitions while retaining essential information. Introduce a hook at the beginning to capture attention, escalate the narrative, and wrap up with a satisfying conclusion. Use techniques like pacing, tone variation, and audio effects to enhance storytelling.

- **Make Your Podcast Flow.** Editing facilitates the smooth flow of your podcast. Seamlessly transition between segments or topics to keep listeners engaged. Eliminate awkward pauses, verbal tics, and filler words for a polished presentation. Please pay attention to the pacing of conversations, ensuring they feel natural and dynamic without rushing or dragging. Smooth transitions between music, interviews, and narration contribute to a cohesive listening experience.
- **Edit Conversations Thoughtfully.** When editing interviews or discussions, be mindful of natural conversational dynamics. Avoid overly aggressive editing that removes the authenticity of the exchange. Maintain the essence of the conversation while trimming off-topic discussions and long pauses that could lead to listener disengagement.
- **Proof and Fact-Check.** Verify facts, statistics, and names mentioned in your podcast to ensure accuracy. Correct any errors or inaccuracies during editing to maintain your podcast's credibility.
- **Preview and Review.** Before finalizing an episode, preview it to ensure a seamless listening experience. Pay attention to transitions, audio quality, and content coherence. This step lets you

catch any last-minute issues before releasing your episode to your audience.

Stage 2: Sound Design

Sound design in podcasting is an artful process that adds depth and engagement to your content. Let's explore the components of sound design and how they can enhance your podcast:

- **Create a Memorable Intro and Outro.** Your podcast's intro and outro are like its signature tune. Craft a memorable and concise introduction that encapsulates the essence of your podcast. A well-designed intro sets the tone and familiarity for your listeners. It should be dynamic, incorporating music and possibly a brief teaser of what's to come. Similarly, the outro should provide a smooth exit, thanking your audience and perhaps teasing the next episode. These segments create a consistent brand identity and a comfortable listening experience.
- **Use Music to Enhance Your Podcast Story.** Music is a powerful tool for evoking emotions and enhancing storytelling. Choose music that complements your content's mood and narrative. Introductory music can grab attention, while background music during discussions or narration can set the ambiance and emotional

tone. Ensure the music doesn't overpower the spoken content; it should be a subtle companion that enriches the listener's experience.

- **Use Sound Effects in Your Podcast Strategically.** Sound effects can add a layer of realism and engagement to your podcast. They can highlight important moments, create transitions, or enhance the atmosphere. For example, a door creak sound effect can transport listeners to a specific location, and a phone ringing sound effect can indicate a change in topic or segment. However, use sound effects sparingly and purposefully, ensuring they contribute to the narrative rather than distracting from it.

- **Source High-Quality Audio.** Ensure that your music and sound effects are high quality and properly licensed. Low-quality or unlicensed audio can detract from the professionalism of your podcast.

Stage 3: Mixing

Mixing combines and enhances audio elements to create a balanced, polished, and professional-sounding podcast. Let's delve into the details of each aspect of mixing:

- **Organize Tracks and Audio Clips.** Begin by organizing your podcast's audio elements on separate tracks. This includes your voice recordings, guest interviews, music, sound effects, and other audio clips. Proper organization makes managing and manipulating each element easier during the mixing process.

- **Improve Your Podcast's Tone Using Equalization.** Equalization (EQ) is a crucial tool in shaping the tonal balance of your podcast. Use EQ to adjust the frequencies of different audio elements. For example, you can enhance vocal clarity by boosting the mid-range frequencies while reducing muddiness or harshness. Be subtle and surgical with EQ adjustments to prevent over-processing and maintain a natural sound.
- **Use a Compressor to Improve Podcast Sound.** Compressors help control the dynamic range of audio, making quiet parts louder and loud parts softer. This evens the overall sound and ensures all elements are audible without sudden volume jumps. Adjust the compressor's threshold, ratio, attack, and release settings to achieve a controlled and consistent sound level.
- **Use Noise Reduction to Give Your Podcast Clarity.** Noise reduction techniques are essential for removing background noise or hiss that may have been captured during recording. Apply noise reduction plugins or tools to mitigate distractions and enhance clarity. However, be cautious not to overdo it, as excessive noise reduction can

introduce artifacts and affect the natural sound of the recording.

- **Balancing Audio Levels.** Achieve a balanced mix by adjusting the volume levels of individual tracks. Ensure that your voice remains the focal point while other elements like music and sound effects complement it without overpowering. A balanced mix creates a comfortable listening experience across different devices and environments.

Stage 4: Mastering

Mastering is the final step in the audio production process that polishes your podcast perfectly. Let's delve into the specifics of mastering and its key components:[5]

- **Listening and Quality Check.** During mastering, you embark on a thorough listening session. This is your chance to meticulously review all the editing, mixing, and sound design work you've done thus far. The goal is to ensure the podcast flows seamlessly, maintaining a consistent narrative and audio quality throughout.
- **Coherence and Flow.** Your podcast should tell a coherent and engaging story at the mastering stage. Listen attentively to identify any inconsistencies, awkward transitions, or abrupt changes that might have been missed during

previous stages. Smooth out these elements to ensure the podcast maintains a captivating flow.

- **Volume Level Adjustments.** One of the key tasks in mastering is ensuring that all your tracks' volume levels are balanced and consistent. Address any noticeable volume disparities between segments, guests, or sound effects. This process helps prevent sudden jumps in volume that could be jarring to listeners.
- **Matching Tracks.** If your podcast includes various audio elements such as interviews, narration, music, and sound effects, the mastering stage is where you fine-tune their balance. Ensure that these components blend harmoniously and none overshadow the others. Achieving a cohesive sonic palette enhances the overall listening experience.
- **Additional EQ Work.** While major equalization adjustments are typically made during the mixing stage, mastering may involve subtle tweaks. Focus on overall tonal balance and address any remaining tonal inconsistencies. Be cautious and make minimal changes to avoid altering the podcast's character.
- **Exporting the Final Master.** Once you're satisfied with the sound and coherence of your podcast, export the final master. Choose a suitable audio format, such as WAV or FLAC, for the best quality. Consider the specific requirements of

podcast platforms and ensure your file size is manageable for online distribution.

PODCAST EDITING TIPS

Podcast editing is a refined craft that elevates your content. Here's a breakdown of essential podcast editing tips to help you achieve a polished and engaging final product:[6]

Listen to Your Interview Once Before Editing

Begin by giving the entire interview a full listen. This helps you understand the flow, dynamics, and potential editing points. It also gives you a sense of the overall tone and direction of the episode.

Take Notes of Things to Edit Using Timestamps

As you listen, jot down notes on areas that require editing. Use timestamps to mark specific spots where you'll need to trim, remove errors, or enhance the content.

Keep the Blank Space While Editing

Don't rush to remove every pause or silence when you're editing. Natural pauses contribute to a conversational feel and allow listeners to absorb information. Over-editing can make conversations sound unnatural and disjointed.

Listen at Regular Speed

Listen to the recording at the regular playback speed while editing. It helps you maintain a natural sense of pacing and ensures that the final product sounds good to your audience.

Editing When One Person Is Quieter

If one person's audio is quieter than the other's, use audio normalization to balance the volume levels. You can also use compression to bring up quieter audio while controlling louder sections. Ensure a consistent volume without introducing distortion.

Editing a Podcast Trailer

Podcast trailers are brief previews that entice listeners. Focus on capturing the essence of your podcast, showcasing its unique selling points. Keep the trailer concise, engaging, and reflective of your podcast's style.

Using Sound Effects Tastefully

Sound effects can enhance storytelling if used judiciously. Use them to punctuate moments, and transitions, or create ambiance. Sound effects should complement the content and not overwhelm it. Maintain a balance between narration, dialogue, and effects.

Give It One Last Listen

After editing, give the entire episode one final listen. This helps catch any overlooked mistakes, ensures a smooth flow, and lets you evaluate the overall quality of the episode.

Choosing Podcast Music

Choose music that aligns with your podcast's tone and theme. It should enhance the listening experience without distracting from the content. Look for royalty-free or licensed music to avoid copyright issues. The intro and outro music should be memorable and recognizable, contributing to your podcast's branding.

BRINGING IT ALL TOGETHER

As you can see, recording, editing, and mastering your podcast is crucial to ensuring a high-quality audio experience. Your listeners will greatly appreciate getting your content free from any hindrances from poor audio, distracting noise, or annoying sound effects.

Effectively recording, editing, and mastering your episodes are a major component of each show's structure. Employing a coherent structure provides your audience with a roadmap they can follow. They can understand your thought process in each episode while knowing what to expect from your content.

Above all, strive to be yourself. While that may sound cliché, the fact is that trying to copy others' work just won't cut it. Audiences pay a premium for authentic content that resonates deeply and personally. That is why the next chapter will look at how you can structure your episodes to bring out the best in your content while engaging your audience to the fullest.

5

STRUCTURING YOUR PODCAST EPISODES

If you fail to plan, then you are planning to fail.

— BENJAMIN FRANKLIN

Crafting a remarkable podcast hinges on the wisdom encapsulated in the adage, "If you fail to plan, then you are planning to fail." This adage holds particular relevance in the world of podcasting. A successful podcast necessitates meticulous planning encompassing various aspects: content, format, target audience, and promotion strategy. Thoroughly defining your podcast's purpose and identifying the interests of your listeners enables you to curate engaging and valuable episodes. Adequate scripting and episode structuring maintain coherence and flow. Strategic scheduling and effective marketing and distribution plans amplify your podcast's reach. Embracing this adage underscores that a well-thought-out plan is the bedrock of a triumphant podcasting journey.

THE IMPORTANCE OF EPISODE STRUCTURE

Maintaining a structured podcast episode ensures a cohesive and engaging listening experience. The risk of appearing disjointed and rambling demands a well-thought-out episode structure. This structure encompasses an engaging introduction, a well-paced middle section, and a compelling closing with a clear call to action.

Begin with a concise and captivating introduction that outlines the episode's theme or topic. This sets the stage for what's to come and hooks the audience's interest. In the middle section, organize your content logically. Break down complex concepts into manageable segments, using clear transitions to guide the listener smoothly through the episode.

As you approach the conclusion, summarize the main points and reiterate the key takeaways. Conclude with a strong call to action, encouraging listeners to engage further, share, or explore related content. So, we're going to delve into the importance of structuring episodes effectively, providing your listeners with the best possible experience.

Podcast Structure Matters[1]

Podcast structure is a foundational aspect that significantly impacts the quality and impact of your content. Without structure, you may quickly lose your audience, especially if your show's point is unclear from the start.

All good narratives demand a sense of cohesion and flow. Speaking without a logical or planned direction can confuse your audience and result in a messy and disjointed conversation. Maintaining a structured approach ensures that your discussions remain coherent, preventing listeners from getting lost and enabling them to stay engaged.

Structure provides a framework for effective storytelling. Contrary to stifling creativity, having predetermined episode parameters serves as a guide, not a restriction. Following a "1,2,3" structure (opening, middle, conclusion) offers production listeners a clear roadmap for where different content elements belong. This clarity enhances storytelling, enabling hosts to craft compelling narratives with purpose and direction.

Additionally, structure eliminates guesswork. Knowing the trajectory of your content, how it contributes to the overall story, and the desired tone simplifies the creative process. This clarity often sparks more innovative ideas that enhance audience engagement. That is why a well-structured podcast elevates the listener experience, making it easier to follow, more enjoyable, and ultimately more successful.

The challenge, however, is knowing how to structure podcast episodes. Following an opening-middle-conclusion structure is highly effective. That is why discussing the nuts of bolts of podcast structuring is crucial to your show's success.

How to Structure Podcast Episodes[2]

There is no magic formula for structuring podcast episodes. Instead, there is a solid framework based on experience and logic. This framework can be adapted based on your show's specific topics and target audience. That is why understanding who your listeners are and what value you deliver will help you tweak this framework to meet your specific needs. Here's a detailed look at how to structure podcast episodes so you can build top-quality, engaging content.[3]

- **Plan for Episode Length.** Determine the ideal length for your episode based on your content and target audience. A clear time frame helps maintain listener interest and expectations.
- **Three Act Structure.** Use the three-act structure (opening, middle, conclusion) to guide your episode's flow. The beginning sets the scene, the

middle develops the content, and the end summarizes the main points.

- **Plan for Pacing.** Balance the pacing of your episode by alternating between informative segments, anecdotes, and discussions. Keep the energy level consistent to maintain listener engagement.
- **Episode Theme.** Define a central theme for your episode. This theme provides coherence and direction, helping you stay focused and on track.
- **Introduction, Back Sell, and Greeting.** Start with a captivating introduction that outlines the episode's theme. After the main content, include a "back sell" to remind listeners of key takeaways and encourage revisiting. Greet your audience with enthusiasm at the beginning.
- **Raise Early Questions.** Pose thought-provoking questions early on to pique listeners' curiosity. These questions serve as guideposts throughout the episode, keeping the conversation purposeful.
- **Promote Other Content.** Strategically integrate mentions of other relevant content, such as previous episodes, website resources, or upcoming events. This promotes engagement and expands your podcast's reach.
- **Closure, Satisfaction, and End.** Reach closure by summarizing the main points and answering any questions raised. Provide a sense of satisfaction to the listeners by delivering on the promises made

in the introduction. Always have the end in mind to avoid abrupt endings.

- **Front Sell, Thank Your Audience.** Before concluding, provide a "front sell" by previewing the next episode or relevant content. Conclude by thanking your audience for their time and attention.
- **Use Templates and Checklists.** Develop templates and checklists for episode planning and production. These tools ensure consistency and help maintain quality across episodes, from scripting to editing.

As you can see, this framework can go a long way toward building a killer episode every time. Most importantly, your audience will see a consistent rationale behind every show.

The Three-Act Format

The "Three-Act" format is widely used among podcasters since it's easy to implement and highly effective. Like any play or movie, the three-act format guarantees a solid foundation for your podcast. Let's take a closer look at this format in action:[4]

Act 1: The Intro

In the first act, the introduction or opening, you set the stage for your podcast episode. This is where you grab your audience's attention and introduce the main topic or theme of the episode. You aim to hook the listeners and give them a reason to keep listening. You might share a captivating anecdote, pose a thought-provoking question, or provide a brief overview of what's to come. This act creates a sense of anticipation and establishes the context for the following content.

Act 2: Body of the Content

In the second act, the body of the content or the middle, you delve deep into the main topic. This is where the bulk of your information, analysis, and discussion occurs. Break down your content into smaller segments or points to maintain clarity and structure. Each segment should build upon the previous one, contributing to the overall narrative. Transition smoothly between segments using clear cues or introductions. This act provides the substance of your episode and keeps the audience engaged with valuable insights, stories, and examples.

Act 3: The Outro/Resolution

The third act, also known as the outro, conclusion, or resolution, brings your episode to a satisfying conclusion. Summarize the main points discussed in the body of the content and offer key takeaways. Address any questions raised earlier in the episode and provide closure by wrapping up loose ends. You can also use this act to tease upcoming content or episodes, keeping your audience excited about what's to come. Express gratitude to your listeners for tuning in. End with a strong call to action, encouraging engagement, feedback, or further exploring related topics.

Keep in mind that the "Three-Act" format offers a natural and effective way to structure your podcast episodes, ensuring a clear beginning, middle, and end. It helps maintain listener engagement by providing an organized and coherent narrative flow, making your content more enjoyable and impactful.

CREATING CAPTIVATING INTRODUCTIONS

Hooking your audience isn't about producing a catchy intro with a snazzy tune. It's about piquing your audience's attention from the start. Thus, creating a positive first impression lets your listeners see the substance in your content from the first moment they tune in. A strong opening grips your audience's attention, leading them to consume your content. In contrast, an ineffective opening can easily turn listeners off, causing them to tune out just as easily as they tuned in.[5]

Be aware that an effective intro doesn't magically happen. It takes careful attention to detail and refinement. That is why you must be prepared to work on it until it is just right. With that in mind, the following points will help you craft a great intro from episode one:[6]

- **Start with the Basics.** Begin by stating the name of your podcast. Make sure it's clear and easy to remember. Mention your name as the host and introduce any guests you have on the episode.
- **Use a Catchy Tagline.** Incorporate a brief tagline that succinctly conveys the essence of your podcast. This tagline should give listeners an idea of what to expect and why they should listen.
- **Add Music and Sound Effects.** Integrate music or sound effects that align with your podcast's tone. This adds a dynamic and engaging element to your introduction, setting the mood and drawing in your audience.
- **Include the Episode Number and Title.** State the episode number and title to give context to your listeners. This helps them keep track of the content they've already listened to and understand what the current episode is about.
- **Define Your Audience.** Clearly establish who your podcast is intended for. Briefly describe your target audience's interests, needs, or pain points that your podcast addresses. This helps listeners self-identify and feel a connection to your content.
- **Mention Sponsors.** If your podcast has sponsors, mention them during the introduction. Keep it concise and relevant to the episode's content. This also helps support your podcast financially.

Remember, brevity is key when crafting podcast introductions. Aim to keep it under a couple of minutes while incorporating these elements. The goal is to spark curiosity, showcase your podcast's unique value, and give listeners a reason to continue listening.

CRAFTING ENGAGING CONTENT AND STORYTELLING

Your show's content is its lifeline. Without it, there would be no show. That is why investing the time and effort to produce compelling content is well worth it. But beyond great content, knowing how to deliver it is key. An ineffective narrative can derail great content, turning audiences away. You'll build a loyal fan base by combining effective storytelling and compelling content. This fanbase will reward your efforts with their time, attention, and great reviews. So, let's discuss how you can craft engaging content and storytelling to take your show to the next level:[7]

- **Follow Your Interests.** Share topics you're passionate about. Your enthusiasm will resonate with listeners and make your content more engaging.
- **Audience-Centric Approach.** Tailor your content to your target audience's interests, needs, and preferences. Understanding their perspective enhances engagement.

- **Harness the Power of Stories.** Stories create emotional connections. Use anecdotes, examples, and narratives to illustrate your points and make your content relatable.
- **Empower Action.** Guide your audience towards actionable steps or insights they can apply. Your content becomes valuable when it helps them progress.
- **Interact with Your Listeners.** Ask questions that encourage audience participation. Incorporate listener feedback and stories to foster a sense of community.
- **Stay Focused.** Stick to the main topic to maintain clarity and coherence. Avoid excessive tangents that might distract from your central message.
- **Feature Unique Experts.** Invite knowledgeable and distinctive experts who can contribute fresh perspectives and insights to your content.
- **Listen Actively.** When interviewing guests, attentively engage with their responses. Meaningful conversations generate deeper connections.
- **Choose Guests Wisely.** Select guests who align with your podcast's goals and resonate with your audience. Diversity of experience and viewpoints can enrich discussions.
- **Optimize Audio Quality.** High-quality audio is crucial for a pleasant listening experience. Invest in good equipment and sound editing for clarity.

- **Provide Next Steps.** Direct your listeners to additional resources, websites, or references related to the episode's content for further exploration.
- **Stay Authentic.** Be genuine and let your personality shine through. Listeners connect with hosts who are authentic and relatable.
- **Transcriptions for Accessibility.** Offer transcriptions of your episodes for accessibility. This enables a broader audience to engage with your content.

Like your show's other aspects, crafting the best possible content takes time and effort. So, don't shy away from refining your content until you find that perfect balance between great storytelling and highly useful content.

PERFECTING YOUR STORYTELLING SKILLS

Developing your storytelling skills takes time and practice. The good news is that the following best practices can help you hit the ground running, greatly reducing the time needed to perfect your content delivery. Let's talk about how to perfect your storytelling skills:[8]

- **Know Your Audience.** Understand your listeners' preferences, interests, and demographics. Craft stories that resonate with them for maximum engagement.

- **Thorough Research.** Gather comprehensive information on your chosen topic. Accurate details lend credibility and enhance the richness of your storytelling.
- **Understand Story Structure.** Learn the classic story structure: introduction, rising action, climax, falling action, and resolution. Organize your content accordingly.
- **Opening, Middle, Conclusion.** Each story should have a clear beginning, where you introduce the setting and characters; a middle, where you build tension and develop the plot; and an end, where you provide resolution.[9]
- **Develop Characters.** Create relatable and compelling characters that your audience can connect with emotionally. Their motivations and struggles enhance your story's depth.
- **Forge Emotional Connections.** Infuse emotions into your storytelling. When listeners empathize with the characters' experiences, they become more invested in the narrative.
- **Craft a Narrative Arc.** Design a journey that takes your audience through various emotional and narrative highs and lows. This keeps them engaged and eager to see how the story unfolds.
- **Every Moment Counts.** Trim any unnecessary details to maintain a compelling pace. Every moment should contribute to the story's progression or character development.

- **Ask Relevant Questions.** If interviewing guests, ask probing questions that reveal insights, anecdotes, and personal experiences. This adds depth to your storytelling.
- **Utilize Sound Effects and Music.** Enhance the atmosphere and mood of your stories with well-chosen sound effects and music. These elements add a layer of immersion.

Keeping these tips in mind will significantly reduce the learning curve, helping you create an amazing show right from the beginning.

BEST PRACTICES FOR CONDUCTING INTERVIEWS AND CO-HOSTED EPISODES

Interviews are a staple of podcasts. Interviews introduce your audience to other great voices, opinions, and ideas. Similarly, adding co-hosts can spice up your show, putting a refreshing spin on your show. Nevertheless, incorporating interviews and co-hosts requires its own structure and approach. Here are some best practices to help you incorporate these elements effectively:[10]

- **Find Guests of Interest.** Select guests who align with your podcast's theme and who genuinely intrigue you. Your passion will translate into engaging content.

- **Research Your Guest.** Familiarize yourself with your guest's background, achievements, and areas of expertise. This knowledge helps you ask relevant and insightful questions.
- **Explore Their About Page.** Check their website or bio for valuable information about their experiences, achievements, and interests. This provides a starting point for your conversation.
- **Review Social Media Profiles.** Examine their social media presence to understand their opinions, hobbies, and engagement with their audience. This informs your approach.
- **Find Other Media Appearances.** Listen to or watch other interviews they've participated in. This helps you avoid asking repetitive questions and adds depth to your conversation.
- **Prepare Probing Questions.** Craft open-ended questions that delve into your guest's experiences, insights, and opinions. These encourage in-depth responses.
- **Pre-Interview Process.** Reach out to your guest to discuss the interview's flow, expectations, and any specific topics you'll cover. This establishes rapport and clarity.
- **Keep the Conversation Moving.** Guide the conversation smoothly by transitioning between topics. Maintain a balance between structure and organic flow.

- **Respectful Conversation.** Avoid interrupting your guest. Let them express their thoughts fully before interjecting with follow-up questions or commentary.
- **Practice Active Listening.** Pay close attention to your guest's responses. Active listening ensures you catch nuances and enables you to ask thoughtful follow-ups.
- **Listen to Your Interviews.** Review your own episodes to identify strengths and areas for improvement. This helps refine your interviewing technique.
- **Learn from Experts.** Study renowned interviewers to understand their styles and techniques. This can inspire fresh approaches to your interviewing.

Be sure to listen to your audience. More often than not, they'll suggest guests they would love to see on your show. Try your best to incorporate these suggestions, as they will keep your audience engaged while building a closer connection.

WRAPPING UP WITH STRONG CONCLUSIONS

A magnificent podcast conclusion is paramount to match your impactful introduction. It's a platform to neatly tie together the ideas you've discussed throughout the episode. Express gratitude to your listeners, encouraging their return for future content or subscribing. A well-crafted call-to-action can further engage your audience, fostering community participation or guiding them toward related resources. Just as a strong beginning sets the tone, a compelling conclusion leaves a lasting impression and reinforces the value of your podcast.

Remember, you must never neglect your show's conclusion. Doing so does you and your audience a huge disservice. That being said, here are the key elements to keep in mind when producing a great podcast conclusion:

- **Invest Time.** Devote effort to crafting your conclusion. It's the final impression you leave on your audience. Just because it's the end doesn't mean it doesn't deserve equal effort and attention to detail.
- **Express Gratitude.** Begin with a heartfelt "thank you" to your listeners for their time and attention.
- **Summarize Key Points.** Recap the main ideas discussed in the episode. This reinforces the takeaways for your audience.
- **Include a Call to Action.** Encourage engagement with a clear call to action—whether it's subscribing, sharing, or participating in a discussion.
- **Preview the Next Episode.** Provide a sneak peek of what's coming next. This keeps your audience excited and engaged.

- **Solicit Feedback.** Invite listeners to share their thoughts, ideas, or suggestions. This fosters a sense of community.
- **Ask for Audience Input.** Encourage your listeners to suggest topics they want to hear about. This enhances audience involvement. Be sure to include a social media handle, email, or messenger address where audiences can reach you.
- **Mention Funding Campaigns.** If applicable, briefly mention ongoing funding campaigns, but keep it non-intrusive.
- **Choose Conclusion Music.** Use music that complements your podcast's tone and sets the mood for the conclusion. This could be a great time to use engaging background music while featuring your theme song (if you have one) as you wrap up the show.
- **Add Easter Eggs.** Include hidden clues or hints related to upcoming content. This can intrigue and excite your dedicated listeners.

Above all, be sure to spark audiences to return for more. Leave open-ended questions and thoughts prompting your audience to come back for the next episode.

BRINGING IT ALL TOGETHER

Having a structured approach to producing podcast episodes is crucial for success. This organized framework provides your audience with a clear and logical flow of content and helps you cultivate a loyal following. Well-structured episodes enable listeners to see the rationale behind your discussions, making engaging and connecting with your content easier.

A structured approach serves as a roadmap for your podcast, ensuring each episode has a purposely defined opening, middle, and conclusion, fostering a coherent progression of ideas. This consistency builds familiarity and trust among your audience.

Moreover, structured episodes are an investment in your podcast's future. They enhance the overall quality of your content, making it more appealing for new listeners and encouraging existing ones to return. This attention to detail pays off in the long run, increasing the likelihood of your podcast gaining traction and achieving success. Crafting successful episodes is crucial, especially considering the investment hosting and distribution require moving forward. In the next chapter, we will discuss how your show can make investing in hosting and distribution a worthwhile endeavor.

6

PODCAST HOSTING AND DISTRIBUTION

You're going to be terrible at a lot of things for years until you're successful. People probably aren't going to listen to your podcast initially. But if you like it and you keep putting it out, people will find it.

— KEITH KINGBAY

R emember, success is a journey, not an instant destination. Don't be disheartened if your podcast doesn't initially get your hoped-for attention. It's completely normal to be less than perfect at first. What truly matters is your passion and persistence. Keep refining your content, honing your skills, and sharing your unique voice with the world. Your dedication will pay off in time, and listeners will discover and appreciate your podcast. So, embrace the learning curve, stay true to your passion, and don't be discouraged by initial setbacks. Your journey to success is paved with every effort you make.

With that in mind, this chapter will focus on hosting and distributing your podcast. This step is a major milestone in building significant traction and wide recognition for your show.

DIFFERENTIATING HOSTING PLATFORMS FROM DIRECTORIES

Setting up a hosting and distribution platform is a key step in helping your podcast gain traction and build a wide following. To begin with, let's define each term.

A hosting platform for your podcast is a service that stores and delivers your podcast episodes to listeners on the internet. Think of it as a digital home for your audio content. These platforms provide the necessary infrastructure to upload, store, and distribute your episodes to popular podcast directories like Apple Podcasts, Spotify, and more. They also offer features like analytics, RSS feeds, and customizable show pages. Essentially, podcast hosting platforms make sharing your content with a global audience easier, track its performance, and ensure it remains accessible to listeners anytime, anywhere.[1]

Conversely, a podcast directory is an online platform or database that lists and organizes podcasts. It is a search engine and directory for users to discover, subscribe to, and listen to podcasts. Popular directories include Apple Podcasts, Spotify, Google Podcasts, and many others. Podcasters submit their show's RSS feed to these directories, allowing their content to be indexed and easily found by listeners. Users can explore podcasts by category, genre, or topic of interest, subscribe to their favorite shows, and listen to episodes through the directory's app or website. Directories play a crucial role in helping podcasts reach a wider audience.[2]

Let's now take a look at the key differences between a hosting platform and a directory:[3]

Podcast Hosting Platforms

- **Purpose.** Hosting platforms are primarily used for storing and managing podcast audio files. After recording, editing, and exporting an MP3 file, podcasters upload their episodes to these platforms for storage and distribution.
- **Functionality.** Hosting platforms serve as online storage for audio files and are responsible for delivering podcast episodes to listeners when they access them through directories or websites. They provide the backend infrastructure for podcast distribution.

- **Accessibility.** Podcast creators typically access podcast hosting platforms for uploading and managing their content. Listeners do not typically interact directly with hosting platforms.

Podcast Directories

- **Purpose.** Podcast directories allow listeners to discover, subscribe to, and consume podcasts. They are where users find new podcasts, browse episodes, and listen to content.
- **Functionality.** Directories aggregate podcast feeds from hosting platforms and display them to users. When listeners choose to play an episode, it streams from the hosting platform. Directories offer a user-friendly interface for exploring and enjoying podcasts.
- **Accessibility.** Podcast directories are the apps and websites listeners commonly use to access and consume podcast episodes. They are the front-end interfaces where users search, subscribe to shows, and listen to episodes.

Podcast hosting platforms are behind-the-scenes tools for podcast creators to store and distribute their audio files. In contrast, podcast directories are user-facing platforms where listeners discover and enjoy podcasts. Both components play crucial roles in the podcast distribution ecosystem, with hosting platforms ensuring file storage and delivery and directories facilitating content discovery and consumption.

PODCAST HOSTING PLATFORMS

Choosing a podcast hosting platform is a major decision in building a successful show. Think of a podcast hosting platform as the specialized storage space for your podcast's audio files, just like a website needs a hosting service to store its content. These platforms provide a secure and reliable home for your podcast episodes. They're equipped with servers optimized to handle large media files like MP3s, ensuring they're readily available for download and distribution to your listeners.

Like choosing a web hosting provider to keep your website accessible online, selecting the right podcast hosting platform ensures that your audio content is efficiently stored, delivered, and accessible to your audience. It's a crucial step in the podcasting journey that ensures your episodes reach your listeners smoothly and reliably.[4]

Why Do You Need a Podcast Hosting Platform?

When you record a podcast, you generate a substantial number of media files, including audio and sometimes even video. While it might be tempting to create your own website to host and share these files, most standard website hosts aren't equipped to handle large media files' storage and bandwidth requirements. This can lead to a less-than-ideal user experience, with issues like slow loading times or, in worst cases, website crashes.

This is where a podcast hosting platform comes to the rescue. Podcast hosting sites are specifically designed to cater to the needs of podcasters. They offer ample storage space, ensuring your media files are safely stored and readily available for as long as you want. Moreover, these platforms often provide tools to assist you in distributing your podcast to popular directories and platforms. This means your podcast will reach a wider audience, making it easier for people to discover and tune in to your show.

A podcast hosting platform simplifies the technical aspects of podcasting, ensuring a seamless experience for you and your audience. It eliminates the worries of running out of storage space or dealing with performance issues on your website. So, opt for a dedicated podcast hosting platform to share your content with the world for the best results and hassle-free podcasting journey.[5]

The Benefits of Utilizing a Podcast Hosting Platform

Podcast hosting platforms offer many benefits that can greatly enhance your podcasting experience. Let's explore these advantages:[6]

- **Robust Analytic.** Hosting platforms provide detailed analytics tools that give valuable insights into your audience's behavior. You can track metrics such as the number of downloads, listener demographics, popular episodes, and geographic locations. These insights help you refine your content strategy, target your audience better, and make data-driven decisions to grow your podcast.

- **Faster Speed.** Podcast hosting platforms are optimized for delivering media content swiftly. They are equipped with high-speed servers and content delivery networks (CDNs) that ensure your episodes load quickly, providing a seamless listening experience for your audience. Slow loading times can deter listeners, so having a hosting platform with fast speeds is crucial.
- **Smoother Data Transfers.** These platforms are designed for efficient data transfer, ensuring your episodes are consistently available to listeners. They manage the technical aspects of file delivery, so you don't have to worry about bandwidth limitations or server maintenance. This reliability means your podcast remains accessible even during traffic spikes.
- **No Compromise in Quality.** You can upload your episodes in their original, high-quality format with dedicated hosting platforms. The platforms handle the necessary encoding and compression for various devices and bandwidths, ensuring your content sounds crisp and clear to all listeners.

- **Repurpose Your Content.** Many podcast hosting platforms offer features for repurposing your podcast content. You can easily create teaser clips, shareable snippets, or even transcriptions, extending the reach of your content across different platforms like social media or your website.

In short, podcast hosting platforms are not just storage spaces for your audio files; they provide a suite of essential tools and advantages that contribute to the success and growth of your podcast.

The Best Podcast Hosting Platforms

When looking for the right hosting platforms for your podcast, the following offer robust features and easy-to-use interfaces. Here's a closer look:[7]

- **Buzzsprout.** Known for its user-friendly interface and ease of use, Buzzsprout offers a range of features, including detailed analytics and distribution options.
- **PodBean.** PodBean is popular for its comprehensive podcast hosting and monetization options, making it a great choice for beginners and seasoned podcasters.
- **Captivate.** Captivate stands out with its advanced analytics, customizable podcast websites, and

user-friendly interface, making it an excellent option for podcasters looking to grow their audience.

- **Transistor**. Transistor offers a professional hosting solution with podcast analytics, multiple podcast management, and an emphasis on team collaboration.
- **Castos**. Castos integrates seamlessly with WordPress and offers one package of podcast analytics, transcription services, and podcast hosting.
- **Resonate**. Resonate focuses on monetization with features like listener support and private podcast hosting.
- **Libsyn (Liberated Syndication)**. As one of the oldest podcast hosting platforms, Libsyn provides reliable hosting and distribution and detailed statistics.
- **SoundCloud**. SoundCloud is known for its music but also hosts podcasts. It's suitable for podcasters looking to tap into its large and diverse user base.
- **Anchor**. Anchor is a free hosting platform owned by Spotify, making creating and distributing podcasts easy. It's beginner-friendly.
- **Audioboom**. Audioboom offers monetization opportunities and content distribution through its network, making it a choice for podcasters seeking wider exposure.

- **RSS.com**. RSS.com provides straightforward podcast hosting with reliable file storage and distribution.
- **Spreaker**. Spreaker offers live podcasting capabilities, monetization options, and various hosting plans to suit different needs.
- **Blubrry**. Blubrry is known for its powerful podcasting plugins for WordPress and offers podcast hosting with analytics and distribution.
- **Simplecast**. Simplecast provides podcast hosting focusing on analytics and marketing tools to help podcasters grow their shows.
- **Fusebox**. While not a hosting platform, Fusebox offers podcasters a customizable audio player and tools for enhancing the listener experience.

Each podcast hosting platform has unique features and strengths, so the choice depends on your specific needs and goals as a podcaster. Explore their offerings thoroughly to find the best fit for your podcasting journey.

PODCAST DIRECTORIES

[8]Did you know that when you listen to a podcast on Apple Podcasts or Spotify, the content isn't stored on these platforms? Podcasts find their home on a podcast host, not in directories. Think of directories as digital phonebooks – they help you find what you're looking for. When you hit play on a podcast in a directory, it's like a bridge

connecting you to the audio stored elsewhere on your podcast host.

For podcasters, directories are the gateway to reaching a wider audience. Some directories specialize in specific niches, connecting you with a highly targeted audience (think NASCAR fans). Others, like Apple Podcasts and Spotify, cover a vast array of topics, making them the go-to spots for podcast discovery. So, directories are your ticket to getting heard and attracting new listeners, whether in a niche or casting a wide net.

Let's now take a close look at the best podcast directories on the market:

Apple Podcasts

Widest Reach: Apple Podcasts is one of the most popular podcast directories globally, making it essential for podcasters looking to reach a broad audience.

User-Friendly: Its user-friendly interface and extensive library make it a go-to choice for new and seasoned podcast listeners.

Spotify

- **Music Integration.** Spotify seamlessly integrates podcasts with its music streaming service, exposing podcasts to millions of music enthusiasts.

- **Discovery Features.** Spotify's recommendation algorithms and curated playlists help listeners discover new podcasts tailored to their interests.

Google Podcasts

- **Google Ecosystem.** Integrated with Google Search and Google Assistant, Google Podcasts ensures your podcast is discoverable through voice commands and web searches.
- **Cross-Platform.** It works across Android and iOS devices, providing access to a wide audience.

Stitcher

- **Premium Content.** Stitcher offers premium content, providing podcasters monetization opportunities through its subscription-based model.
- **Personalized Playlists.** Listeners can create playlists, making organizing and consuming their favorite podcasts easier.

TuneIn

- **Live Streaming**. TuneIn allows live streaming of some podcasts and radio stations, offering real-time engagement with your audience.
- **Podcast and Radio Hub**. It serves as a hub for podcasts and live radio, broadening your podcast's potential reach.

These directories play a crucial role in helping podcasters connect with their audiences. Whether looking for wide-reaching platforms like Apple Podcasts and Spotify or specialized features like live streaming on TuneIn, these directories offer unique advantages to podcast creators and listeners.

Submitting Your Podcast to a Directory

Getting your podcast on a directory is straightforward when you have the right tools in place. Here's what you need:[9]

- **MP3 File**. First and foremost, you'll need an MP3 file of your podcast recording. This is the audio content of your episode, which your audience will listen to.
- **RSS Feed**. Your podcast hosting service will generate an RSS (Really Simple Syndication) feed for your podcast. Think of this as the heart of

your podcast's information. It contains details like episode titles, author names, descriptions, and more. Your hosting service automatically updates this feed with new content and metadata whenever you release a new episode.

Once you have these elements ready, you can submit your podcast to a directory. Most hosting services offer seamless integration, allowing you to do this directly from your dashboard. Simply follow the directory's submission process, usually by providing your podcast's RSS feed URL, and the directory will automatically pull in your episodes.

After the initial setup, the directory will keep track of your podcast and automatically add any new episodes you release in the future.

Follow these steps to ensure your podcast is up and running on your chosen directory:

Create a Podcast RSS Feed on Your Hosting Platform

- Begin by selecting a reliable podcast hosting platform that suits your needs.
- Upload your podcast episodes, including the audio files, titles, descriptions, and other metadata.
- Your hosting platform will automatically generate an RSS feed for your podcast. This feed is the

central hub of your podcast's information, containing details about each episode.

Submit the RSS Feed to Your Podcast Directory

- Choose the podcast directories where you want your podcast to be available. Popular options include Apple Podcasts, Spotify, Google Podcasts, and more.
- Visit the submission page of each directory. You'll typically need to provide your podcast's RSS feed URL, which you obtained from your hosting platform.
- Follow the submission guidelines and complete any required information, such as your podcast's category and artwork.

Wait for Your RSS Feed to Be Approved

- After submission, the directory will review your podcast's RSS feed. This process can take some time, ranging from a few hours to a few days.
- Be patient while the directory checks your feed for compliance with their guidelines. Ensure your podcast meets its content standards and specifications.

Publish and View Your Podcasts

- Once your RSS feed is approved, your podcast episodes will become accessible to a global audience through the directory.
- Listeners can discover, subscribe to, and enjoy your podcast on their preferred directory or app.
- Regularly publish new episodes to keep your audience engaged and informed.

Getting Your Podcast on iTunes or Spotify

iTunes and Spotify are places for anyone looking to make a name for themselves in the podcasting world. So, let's explore how you can get your foot in the door on these significant podcast directories:

iTunes[10]

- **Prepare Your RSS Feed.** Ensure your podcast has an RSS 2.0 conforming feed. Most podcast hosting platforms generate this for you. Use Apple's recommended podcast validator, Podbase, to check for compliance and fix errors.
- **Have at least 3 Episodes Ready.** To increase your chances of being featured by Apple, it's recommended to have a minimum of three episodes published before submission.

- **Create a New Apple ID.** Set up a dedicated Apple ID specifically for your podcast. This helps keep your podcast-related activities separate from personal ones.
- **Log into Podcast Connect.** Visit Apple's Podcast Connect platform and log in with your new Apple ID.
- **Add Your Podcast Show.** Click the "+" button to add your podcast show.
- **Select "Add a Show with an RSS Feed."** Choose this option to use your existing RSS feed.
- **Paste Your RSS URL.** On the next page, you'll see a box to paste your podcast's RSS feed URL.
- **Validate Your URL.** Click the "validate" button to check your URL for any issues. You'll see an error message if there's a problem with the feed.
- **Submit Your Podcast.** Once your RSS feed is validated and error-free, click "submit" to send your podcast to Apple for review.
- **Wait for Approval.** The approval process can vary, ranging from a few hours to weeks. Be patient while Apple reviews your podcast.

Spotify

- **Sign up for an Anchor Account or Log In.** If you're new to Anchor, sign up for an account. If you already have one, log in.

- **Create a New Episode.** Click 'Let's do it' in your welcome dashboard to start creating a new episode.
- **Record or Upload Your Episode.** You have two options: record a podcast directly on Anchor or upload a pre-existing recording. Select 'Save Episode' when you're finished.
- **Fill in the Episode Details.** Provide all the necessary episode details, including the title, description, and episode number. Click 'Next' when you're done.
- **Set Up Your Show.** A pop-up will appear, allowing you to 'Set up your show.' Enter your podcast show's details, including the title, description, and other relevant information. Customize it with appealing cover art that represents your podcast.
- **Publish Your Episode.** Once you've filled in all the required information, click 'Publish' or 'Submit,' depending on the platform. Your podcast should automatically appear on Spotify as well.

GUIDELINES TO CONSIDER WHEN SUBMITTING YOUR PODCAST

Here are some crucial guidelines to consider when submitting your podcast to iTunes or Spotify:[11]

Audio Format

- Your podcast audio should be in the ISO/IEC 11172-3 MPEG-1 Part 3 (MP3) format.
- Ensure the bitrate of your MP3 audio falls between 96kbps and 320kbps. This range provides good audio quality for your listeners.

File Size

Keep your episode file sizes below 200MB. This equates to approximately 83 minutes at 320kbps or 200 minutes at 128kbps. Smaller file sizes are easier to manage and download.[12]

Artwork

- Your podcast artwork should be in JPG or PNG format.
- Use a square aspect ratio of 1:1 for your artwork. Square images work well for various podcast platforms and devices.
- Ensure the artwork has the highest resolution for a polished and professional look.

Feed Values

- Limit your episode titles to a maximum of 20 characters. Concise titles are more reader-friendly.
- If you need to include special characters like !@£ $%^&*(), HTML encodes them for compatibility with different platforms and browsers.

Feed Fields

To submit your podcast, it must include the following essential elements:

- **A podcast title.** This is the name of your podcast.
- **Artwork.** Your podcast's square artwork in JPG or PNG format.
- **At least one episode.** Your podcast feed must have at least one episode to be considered for submission.

By adhering to these guidelines, you'll ensure your podcast meets the technical requirements of podcast directories and enhance its overall presentation and accessibility for your audience.

OPTIMIZING METADATA AND DESCRIPTIONS

Think of metadata in podcasting as the "front of the house." It's like your podcast's welcoming porch, the inviting front door, or the appealing front part of your garden. In essence, it's the first impression your podcast makes on potential listeners.[13]

This metadata includes crucial elements that potential listeners see before hitting or reading that play button. It encompasses:

- **Episode Title.** The catchy, informative title piques curiosity and tells listeners about the episode.
- **Episode Description.** A brief but compelling summary of the episode's content entices listeners to explore further.
- **Show Name.** Your podcast's unique name sets the tone and identity for your entire show.
- **Podcaster Details.** Information about you, the podcast host or hosts, giving listeners a glimpse of who's behind the mic.

Now, let's consider the following points to help you optimize your show's metadata, leading audiences to readily find your content available while ensuring a high-quality design and description:

Metadata for Search Engine Robots[14]

- **Cover Art**. Start with compelling cover art. Ensure it's high-quality, eye-catching, and follows the recommended specifications for each podcast directory.
- **Author**. Use your name or your podcasting persona consistently across all episodes. This helps search engines attribute content to you.
- **Show Title**. Make your show title clear, concise, and reflective of your podcast's content.
- Show Description. Craft a compelling show description that provides an overview of what your podcast offers. Include relevant keywords naturally.
- **Episode Title**. Episode titles should be descriptive, intriguing, and relevant keywords to attract listeners.
- **Episode Description**. Write informative episode descriptions that capture the essence of each episode and incorporate keywords when relevant.
- **Tags**. Use relevant tags to help search engines categorize your content correctly. These should align with your podcast's topic and content.
- **Category**. Select the most appropriate category for your podcast. This helps it appear in relevant search results.
- **Language**. Specify the language of your podcast to ensure it reaches the right audience.

- **Explicit**. Indicate if your content contains explicit language or themes, as some listeners may prefer or require content warnings.
- **Website Permalink**. Include a link to your podcast's website or landing page for additional information.

Metadata for Humans

- Craft metadata with your target audience in mind. Make it engaging, informative, and appealing to entice listeners.

General Metadata Optimization Practices

- **Adjust Per Directory if Needed.** Tailor your metadata for different podcast directories if necessary. Some platforms may have specific guidelines or character limits.
- **Never Stop Optimizing.** Continuously assess and refine your metadata based on listener feedback and changing trends in your podcast's niche. Stay adaptable and open to improvements.[15]

BRINGING IT ALL TOGETHER

Getting your podcast on the right platform is pivotal for podcast success. It's like choosing the perfect location for your storefront; visibility matters. Your podcast's platform sets the stage for its discoverability and accessibility. Once you've secured your spot, effective marketing and promotion strategies can work their magic. You want your content to shine where your target audience gathers. So, whether it's Apple Podcasts, Spotify, or other platforms, ensure you're in the right place to reach your audience, making your marketing efforts more potent.

Remember, choosing the right hosting platform and directory is crucial since the next step, marketing and promotion, builds on the solid foundation you have laid in crafting a professional-quality podcast.

7

EFFECTIVE MARKETING AND PROMOTION STRATEGIES

You just need one person to listen, get your message, and pass it on to someone else. And, you've doubled your audience.

— ROBERT GARRISH

It's incredible how a single attentive listener can amplify your message. Imagine sharing your thoughts or ideas with one person, and they, in turn, pass it along to another. It's like a ripple effect, doubling your audience with each step. This simple communication can have a powerful impact, spreading your message far and wide. So, never underestimate the influence of someone who takes the time to understand and share your message honestly.

Whether it's a personal story or a grand vision, all it takes is that initial connection, and your message can touch the hearts and minds of many. That is why effective marketing and promotion strategies focus on connecting individuals. When this occurs, your podcast has a unique opportunity to become a force to be reckoned with. Let's jump in.

BUILDING A STRONG BRAND FOR YOUR PODCAST

Podcast branding is a savvy marketing approach that breathes life into your business by infusing it with a distinct visual identity and voice. It's all about showcasing your brand's unique personality and messaging in an engaging auditory format. What sets podcast brands apart is their commitment to aligning their mission with content free from intrusive advertisements. By skillfully integrating visuals, content, and structure, a robust brand strategy delivers a compelling message and captivates a specific audience. This creates awareness and a loyal following that eagerly anticipates each new episode.[1]

So, if you want to elevate your business's presence and storytelling prowess, consider leveraging the power that podcast branding offers shows with a singular value proposition. Here's a deep look at how you can leverage this power:[2]

- **Do Your Homework.** Before launching your podcast, research your target audience. Understand their preferences, interests, and the kind of content they engage with. Analyze your competitors' podcasts to identify gaps and opportunities in the market. This research will help you tailor your content to meet the needs of your audience.
- **Understand Your 'Why'.** Clarify your podcast's purpose or "why." Why are you starting this podcast? What message or value are you aiming to deliver? Ensure your podcast aligns with your brand's mission and resonates with your audience.
- **Have A Game Plan.** Map out a content strategy in advance. Determine the themes, topics, and format of your episodes. A clear plan will keep your podcast organized, consistent, and engaging.
- **Prepare A Few Months' Worth of Podcasts.** It's a good idea to have several episodes ready before your launch. This content buffer ensures you maintain a regular posting schedule, even if unexpected issues arise.
- **Start True, Build True, And Market True.** Stay true to your brand's identity throughout your podcast journey. Consistency in your message, tone, and values is crucial for building brand trust and recognition. Additionally, authentically market your podcast to attract listeners who resonate with your brand.

- **Don't Stick Too Rigidly to An Outline**. While planning is essential, don't be afraid to go off-script occasionally. Engaging conversations and spontaneous moments can add depth to your podcast and make it more relatable to your audience.

- **Post New Episodes Every Other Day**. Posting frequency is important, but daily episodes can be challenging to sustain. Consider a posting schedule that suits your capacity and audience expectations. Consistency is key.

- **Show What Your Brand Stands For**. Infuse your brand's values and beliefs into your podcast content. Whether it's environmental sustainability, social justice, or innovation, make sure your podcast reflects what your brand stands for.

- **Focus On Delivering Value to Your Listeners**. Your audience is the heart of your podcast's success. Prioritize delivering content that educates, entertains, or informs them. Providing value is the foundation of building a loyal listener base.

- **Broadcast Around Your Brand Values**. Align your podcast content with your brand's core values and principles. This alignment reinforces your brand's identity and attracts like-minded listeners who share your values.

Remember, the golden rule is to stay true to your values. These values are what draw listeners to your show. Avoid compromising on your brand identity, especially in the face of corporate advertising. Your show is much more valuable than any single corporate sponsor.

BUILDING A COMMUNITY OF LOYAL FOLLOWERS

Creating a devoted listener community and enthusiastic word-of-mouth supporters is paramount for every podcast. To achieve this, understanding the unique journey of your podcast audience is crucial.[3] Think of it as the Podcast Listener Journey framework, akin to a typical customer's path through the purchase journey and sales funnel. It encompasses the awareness, consideration, and decision stages but with a few podcast-specific twists. By comprehending this journey, you can effectively tailor your content and engagement strategies, ensuring your podcast resonates at each stage and fosters a strong and dedicated following eager to spread the word about your show.

The Podcast Listener Journey Framework consists of three separate stages. Let's break each one into detail:[4]

Stage 1: Awareness

Firstly, it's crucial to understand how potential listeners become aware of your podcast. A combination of activities and tactics, including effective podcast marketing strategies, often fuels this awareness. Outreach efforts and nurturing your audience are vital, as does featuring guest experts to broaden your reach. Don't underestimate the power of crossover episodes, and consider leveraging paid media to expand your podcast's visibility. Above all, remember that creating genuinely valuable content is the cornerstone of building a loyal listener base. Please remember to stay true to your brand identity, especially as you build a loyal following.

Stage 2: Consideration

The second stage involves establishing your influence and credibility, where trust is critical. Personal interest sparks when your content resonates deeply with your audience. Writing succinctly ensures your message hits home. Remember, first impressions matter, so invest in creating a compelling podcast cover, catchy title, and captivating episode descriptions. Also, don't forget to build a trailer. It's like a sneak peek that can hook potential listeners. Specifically, your trailer should showcase clips and snippets that accurately represent your show's branding, enticing listeners to try your show. Once people check out your show, you have a real shot at building a dedicated base.

Stage 3: Decision to Listen

The third stage refers to people's decision to listen to your show. This decision is based on perceived worth, where nurturing your podcast audience is vital. Make your content valuable and relevant to keep listeners returning for more, fostering the value listeners see in your content. Next, consider the quantity and quality of social proof. Positive reviews, ratings, and testimonials from satisfied listeners can significantly boost your podcast's credibility. Ensuring ease of consumption by providing precise and user-friendly access to your episodes is necessary. Additionally, joining an online community can create a sense of belonging for your audience, fostering deeper connections and conversations.

This framework is a great way to visualize the "sales" cycle in which your show draws new listeners. Like sales conversions, your marketing efforts should be enough to get people through the door. Once inside the shop, your value ought to be enough to knock their socks off, leading to an ironclad online community of which people are proud to be members.

HOW TO BUILD A LISTENER COMMUNITY

Building a listener community is like building a wall. The first bricks at the bottom are the foundation that supports more bricks. As the wall grows, the base becomes more important. That is why building a solid foundation while attracting new listeners is pivotal in building a thriving listener community. Focus even more on your founding members, especially as your show's reach grows. Here are the top tips to help you build a listener community:[5]

- **Answering emails, messages, and DMs.** Interacting with your audience on a personal level is crucial. Responding to emails, messages, and direct messages (DMs) shows that you care and encourages listener engagement and loyalty.
- **Featuring listener questions in episodes.** Incorporating listener questions or feedback in your episodes makes your audience feel heard. It adds an interactive element to your content. It's a great way to create a two-way conversation.
- **Hosting 'Ask Me Anything' sessions.** Periodically hosting AMA sessions allows your audience to ask you questions directly. It's an excellent opportunity to deepen your connection with listeners and share insights or behind-the-scenes stories.
- **Live streaming podcast episodes.** Live streaming adds an exciting and real-time dimension to your

podcast. You can engage with your audience in real time, answer questions, and get immediate feedback.

- **Organizing live in-person podcast events.** Hosting in-person events brings your podcast community together in a tangible way. It allows listeners to meet you and fellow fans, fostering a sense of belonging and community.

Don't forget that your founding members are the key to your show's success. Without them, your show cannot hope to grow in reach. So, reward them by building a "founding member's community" to show your appreciation for believing in you, especially during the early days.

CREATING COMPELLING ARTWORK AND DESCRIPTIONS

[6]The artwork you choose for your show's cover is an essential marketing tactic. Your cover art must reflect your show's personality while drawing new listeners in. Therefore, your cover art is pivotal in establishing your brand identity.

The following considerations are crucial in building the right graphics for your show's cover art:

- Your cover art should visually represent the essence of your podcast. Use imagery, symbols, or text to convey the theme or subject matter immediately. A clear and relevant depiction helps potential listeners understand what to expect.

- Tailor your cover art to resonate with your target audience. Consider their demographics, interests, and preferences. For instance, if your podcast is for tech enthusiasts, your cover art might feature tech-related imagery or a futuristic design.

- The mood and tone of your podcast should be reflected in your cover art. Use vibrant colors and playful graphics if your podcast is light-hearted and humorous. For a more serious or educational tone, opt for a more subdued and professional design.

- Your cover art will appear on various platforms, including podcast directories, social media, and mobile devices. Ensure it's visually appealing and legible in different sizes. Simple and clean designs often work best for versatility.

- Make it clear that your artwork is for a podcast by incorporating recognizable podcast icons, such as a microphone or headphones. Choose a style that aligns with your brand and makes your podcast stand out. Consistency in style across episodes helps with brand recognition.

Remember that less is more. So, keeping your cover art clean and simple is a great way to get your message across effectively. Cluttered cover art can be confusing, causing listeners to miss the point of your show's content and identity.

Additionally, keep in mind that some limitations apply to your cover art. Here is a look:

- Depending on the hosting platform or directory, your show's cover art must be between 1400 and 3000 pixels square.
- Your art should contain 72 dots per inch.
- Use an RGB color.
- The files should be JPEG or PNG images.

Tips on How to Create Effective Podcast Cover Art

Given the importance of your show's cover art, the following considerations will help you create truly compelling graphics:

- [7]**Let Your Content Lead the Way.** Align your cover art with the theme and content of your podcast.
- **Consider Your Audience.** Design with your target audience's preferences and interests in mind.
- Meet Directories' Requirements. Ensure your cover art complies with podcast directories' size and format specifications.

- **Optimize for Various Sizes and Settings.** Make sure your cover looks good in different sizes and on various platforms.
- **Limit Word Count.** Keep text concise and avoid cluttering the design with too many words.
- **Avoid Explicit Content.** Steer clear of explicit language or imagery to maintain a broad appeal.
- **Leave Margin.** Include a margin to prevent important elements from being cut off in various displays.
- **Utilize Colors.** Choose colors that reflect your podcast's identity and message.
- **Consider Color Temperature.** Understand how color temperature can influence the mood of your cover.
- **Typography Matters.** Select typography that effectively conveys your podcast's subject, tone, and style.
- **Think About Dark Mode.** Ensure your cover looks good in both light and dark modes.
- **Don't Pursue Perfection.** While striving for a great design is important, don't let perfectionism delay your podcast launch. Use templates, maintain consistency, opt for high-resolution images, leverage color strategically, compress the final image for web use, and repurpose your artwork to maximize its impact.

Keep in mind that your show's cover art is an evolving entity. So, don't be afraid to tinker with it until you find the right combinations that accurately depict your show's identity and target audience.

PODCAST DESCRIPTIONS

A podcast description, also known as a show description or show summary, serves as the introduction to your podcast for potential listeners. Essentially, it's a brief and concise text that outlines your podcast's core theme or subject matter and provides a glimpse of what listeners can anticipate when they decide to tune in to your episodes. This description plays a crucial role in attracting new listeners, as it serves as the first point of contact and can influence someone's decision to explore your content. Therefore, crafting an engaging and informative podcast description is essential for effectively conveying your podcast's value and enticing potential audience members to listen.[8]

The Difference Between a Podcast Description and Podcast Show Notes

The key difference between a podcast description and podcast show notes is their scope and purpose. Here's a detailed look:[9]

Podcast Description

This is a concise summary of your entire podcast. It offers a broad overview, describing your podcast series's overarching theme, focus, and general content. It's typically static and doesn't change with each episode. The podcast description is like the "About" section of your podcast. It provides potential listeners with a big-picture understanding of what your podcast is about.

Podcast Show Notes

On the other hand, show notes are specific to individual podcast episodes. They accompany each episode and provide detailed information relevant to that particular installment. Show notes typically include episode-specific details such as the title, episode number, mentions of sponsors, a breakdown of topics discussed in that episode, and the names of any guests featured. Show notes are dynamic and change with each new episode, serving as a valuable resource for listeners who want more information about a specific episode's content.

Elements of a Podcast Description

When crafting your show's description, keep the following points in mind:[10]

- **Identify Your Target Audience.** Start by defining who your podcast is intended for. Consider demographics, interests, and needs. State who will benefit from your content.

- **Set Expectations.** Describe what a listener can expect from your show. Mention the overarching theme or subject matter of your podcast. Make it clear what kind of content you'll be delivering.
- **Explain the Value.** Convey the benefits a listener will gain from tuning in. Explain how your podcast will educate, entertain, inspire, or solve problems for your audience.
- **Introduce Yourself.** Provide a brief introduction to yourself or your team if you're hosting the podcast. Mention your qualifications, expertise, or unique perspective that makes you a credible source for the podcast's topic.

Allowing these elements to stand out in your show's description gives listeners the information they need to judge your show's identity, leading them to try your content. So, ensure your show's description truly reflects yourself and your show's identity.

Elements of a Great Podcast Description

So, what makes a podcast description truly captivating? The following points explain the elements of a truly great podcast description:

- **Clear Explanation.** Your podcast description should clearly and concisely explain what your podcast is about. It should define your podcast's overarching theme, subject matter, or focus, ensuring that potential listeners immediately understand what they can expect.

- **Searchable Content.** Make your podcast summary easily searchable by incorporating relevant keywords and phrases. This helps your podcast appear in search results when users seek content related to your topic or niche.
- **Conciseness.** Keep the description concise and to the point. Avoid unnecessary jargon or overly long sentences. A brief but informative description will likely hold the reader's attention.
- **Engagement.** Engagingly craft the description. Use language that captivates the reader's interest and piques their curiosity. Share the excitement and passion you have for your podcast's subject.
- **Conviction.** Convince potential listeners why they should choose your podcast. Highlight the unique value, insights, or entertainment your podcast offers. Explain why your content is worth their time.

A good podcast description effectively communicates the podcast's content, is optimized for searchability, remains concise, engages the reader, and convinces them that your podcast is worth their attention.

Why Podcast Descriptions Are Important

Crafting a compelling podcast description is a secret weapon in podcasting. It's your chance to grab potential listeners' attention as they browse countless options. Think of it as your podcast's elevator pitch—it needs to be engaging, informative, and concise. But it's not just about first impressions; it's also a powerful tool for search engine optimization (SEO). Including relevant keywords and phrases in your description increases the chances of your podcast appearing in search results. So, whether you aim to attract new listeners or boost your podcast's discoverability, a well-crafted description is your ticket to success.[11]

LEVERAGING SOCIAL MEDIA FOR EFFECTIVE STRATEGIES FOR PODCAST PROMOTION

Leveraging social media is an essential part of effectively marketing your show. The following strategies can be a gateway to highly engaging marketing:[12]

Create Shareable Podcast Snippets

Crafting shareable podcast snippets is a must for effective social media promotion. These bite-sized content pieces, whether short audio clips, engaging video teasers, or visually appealing audiograms, serve as potent hooks to reel in your audience. Short audio clips offer glimpses of your podcast's highlights, video teasers blend captivating visuals with audio intrigue, and audiograms provide an enticing visual and audio combo.

Develop Visually Appealing Content

Utilize eye-catching infographics to convey key insights or statistics from your episodes. Share quote cards featuring memorable podcast moments or thought-provoking quotes to captivate your audience. Don't forget to provide glimpses behind the curtain with behind-the-scenes photos and videos, letting your listeners connect with your podcast more personally. These visual elements enhance engagement and make your podcast more share-able, drawing in a wider audience and strengthening your podcast's online presence.

Engage with Your Audience

Take the time to respond to comments and messages promptly, fostering a sense of community and connection. Encourage user-generated content by asking listeners to share their thoughts, questions, or experiences related to your podcast. Hosting Q&A sessions and live events provides valuable interaction and a platform for real-time engagement, where you can directly address your audience's inquiries and build a loyal following.

Leverage Hashtags and Keywords

Harnessing the power of hashtags and keywords is a savvy strategy for promoting your podcast on social media. Research relevant hashtags that align with your podcast's content and audience. Implement keywords strategically in your social media posts and descriptions to enhance discoverability. Consider creating your branded hashtag to foster community and encourage listeners to engage with your content.

Collaborate with Influencers and Guests

Collaborating with influencers and guests can significantly boost your podcast promotion on social media. Cross-promote with your podcast guests by sharing their episodes and tagging them, encouraging them to do the same. Partnering with influencers in your niche can expand your reach to their dedicated followers, bringing new listeners to your podcast. Conduct interviews and takeovers with relevant experts or influencers to introduce your content to their audience and spark curiosity.

Utilize Various Social Media Platforms

To effectively promote your podcast on social media, it's crucial to utilize various platforms wisely. Take the time to understand each platform's unique features and nuances, whether it's Instagram, Twitter, or Facebook.

Tailor your content to cater to the preferences and habits of each platform's audience. Stay active and maintain a consistent presence across all platforms to keep your audience engaged and informed.

Host Giveaways and Contests

Hosting giveaways and contests can be a fun and effective way to promote your podcast on social media. Encourage your audience to participate by sharing and tagging friends, amplifying your podcast's reach. Consider partnering with other podcasters or brands to expand the prize pool and reach a broader audience. Offering exclusive prizes for your podcast's theme or content can create excitement and engagement.

Share Behind-the-Scenes and Personal Stories

Sharing behind-the-scenes and personal stories on social media can help humanize your podcast and connect with your audience more deeply. Showcasing your podcasting process, from brainstorming to recording, provides transparency and insight into your creative journey. Sharing the inspiration behind your episodes lets listeners in on your thought process, making your content more relatable.

Monitor Analytics and Adjust Your Strategy

Monitoring analytics is a crucial aspect of promoting your podcast on social media. Keep a close eye on engagement and growth metrics to gauge the effectiveness of your campaigns. Identify which content types and topics resonate best with your audience and drive the most interaction. By regularly analyzing this data, you can make informed decisions to optimize your social media strategy continuously.[13]

COLLABORATING WITH OTHER PODCASTERS AND INFLUENCERS

[14]Collaborating with other podcasters and influencers can be a mutually beneficial way of boosting a show's audience and reach. Here are highly useful points to leverage the power of collaboration:[15]

- **Find other creators your audience will like**. Identify podcasters or creators whose content aligns with your niche and will likely resonate with your audience. Collaborating with them can introduce your podcast to a relevant and interested audience.
- **Organize "shoutouts" with other podcasters**. Collaborate with fellow podcasters to give each other shoutouts or mentions in your episodes.

This cross-promotion can help you tap into each other's listener base.

- **Appear as a guest on lots of shows**. Offer to be a guest on other podcasts that share your target audience. This lets you showcase your expertise and attract new listeners interested in your subject matter.
- **Podcast as a guest host**. Temporarily host or co-host episodes on other podcasts. This approach can expose you to a new audience and build your credibility in your niche.
- **Use a podcast guest matching service**. Consider using platforms or services that connect podcast hosts with potential guests. These services streamline the collaboration process by matching you with relevant shows.
- **Send a quality podcast collaboration pitch**. When reaching out to potential collaborators, craft a compelling pitch that explains why your podcast would be a valuable addition to their show. Highlight your unique perspective and expertise.
- **Make yourself an amazing collaborator**. Be professional, reliable, and easy to work with. Show respect for your collaborators' time and effort, and maintain a positive and cooperative attitude throughout the collaboration.
- **Create your promotional materials**. Design promotional materials like graphics, banners, or

social media posts to share when collaborating. These materials can help cross-promote your episodes and enhance your podcast's visibility.

Collaborating with other hosts can expand your podcast's reach, connect with a broader audience, and build valuable relationships within the podcasting community, ultimately enhancing your podcast's success.

USING SEO AND KEYWORDS TO OPTIMIZE DISCOVERABILITY

Boosting your show's discoverability is a great way to give it that extra boost it needs to find new listeners. Let's dive into how you can leverage effective tactics to make your show more accessible to new audiences:

What is Podcast SEO?

Podcast SEO, or Search Engine Optimization for podcasts, enhances your podcast's visibility in search engine results like Google. It involves techniques to improve the quality and quantity of traffic to your podcast. By optimizing your podcast content and metadata, you increase the chances of potential listeners discovering your show when conducting online searches. Unlike relying solely on podcast apps for discovery, podcast SEO offers a sustainable method for growing your audience and elevating your online presence. It's a valuable strategy to ensure your podcast reaches a broader audience and gets noticed in the competitive digital landscape.[16]

Podcast SEO Best Practices

Podcast SEO best practices are essential for boosting your podcast's discoverability and attracting a larger audience. Here are the top podcast SEO best practices:[17]

- **Claim Your Podcast on Google Podcasts Manager.** Registering your podcast with Google Podcasts Manager ensures that it's indexed by Google, making it more likely to appear in search results and Google Podcasts.
- **Research Relevant Podcast Keywords.** Identify keywords related to your podcast's content and use them strategically in your titles, descriptions, and episode summaries to improve search engine rankings.
- **Build Authority with Relevant and Quality Content**. Consistently produce high-quality, informative, and engaging podcast episodes that establish your authority in your niche, attracting listeners and search engines.
- **Leverage Podcast Metadata**. Optimize your podcast metadata, including titles, descriptions, and tags, using relevant keywords to improve your podcast's visibility in search results.

- **Increase Your Online Presence with Social Media**. Promote your podcast on social media platforms to expand your reach, increase brand recognition, and drive traffic to your podcast.
- **Create Episode Transcriptions**. Transcribe your podcast episodes to make them more accessible to search engines and those with hearing impairments, improving SEO and audience accessibility.
- **Repurpose Your Podcast Content**. Repurpose your podcast episodes into blog posts, videos, or infographics to reach a broader audience and improve your website's SEO.
- **Build Trust for Your Podcast with Backlinks**. Secure backlinks from reputable websites in your niche. These backlinks signal trustworthiness to search engines and can improve your podcast's search ranking.
- **Work On Your Internal Linking**. Use internal links within your website to connect related podcast episodes and content. This helps improve user experience and keeps visitors engaged longer.
- **Pay Attention to Your Podcast Website Speed and Responsiveness**. Ensure your podcast website loads quickly and is mobile-responsive. Google prioritizes websites with good user experiences, impacting your SEO rankings.

BRINGING IT ALL TOGETHER

Marketing your podcast is a cornerstone of its success, laying the foundation for a devoted follower base. This engaged audience is the key to monetizing your show effectively, as they represent the core demographic that sponsors and advertisers seek. With a solid marketing strategy, you attract more listeners, build trust, and establish your podcast's credibility. Over time, this dedicated community becomes a valuable asset, opening doors to various monetization avenues such as sponsorships, merchandise sales, premium content subscriptions, and more. In essence, effective marketing ensures your podcast's growth. It paves the way for you to turn your passion into a sustainable income stream.

8

MONETIZING YOUR PODCAST

Podcasting is the most convenient media for learning, getting inspired, or being entertained. You can listen to whatever you want, whenever you want, and however you want.

— JENNIFER HENCZEL

I n today's fast-paced world, podcasting is the ultimate beacon of convenience in media consumption. It has transformed how we learn, find inspiration, and seek entertainment. With a diverse array of content available at our fingertips, podcasting provides an unparalleled level of flexibility, allowing us to tailor our listening experience to our unique preferences.

The beauty of podcasting lies in its accessibility. Whether commuting to work, exercising at the gym, or simply relaxing at home, podcasts are there to accompany you, seamlessly integrating into your daily routine. This accessibility transcends geographical boundaries, enabling creators to reach audiences all over the world and fostering global connections based on shared interests and ideas.

In this digital age, where time and space are increasingly compressed, podcasting empowers individuals to curate their own intellectual and emotional journeys, breaking down barriers and making knowledge, inspiration, and entertainment more accessible than ever before.

So, we're going to discuss how podcasting is a highly convenient form of communication and how it can become a steady income source.

EXPLORING DIFFERENT MONETIZATION METHODS

Monetizing your podcast serves as a tangible sign of success, validating the time and effort invested in crafting compelling content. It not only brings financial rewards but also signifies the engagement and loyalty of your audience. When sponsors or advertisers are eager to collaborate, it reflects their trust in your podcast's influence.[1]

Moreover, monetization enables creators to reinvest in their shows, improving production quality and expanding reach. It's a testament to the value you provide to your listeners. It signifies that your podcast has found its voice and a place in the competitive world of digital media, ultimately fueling further growth and innovation.[2]

Find Sponsorship Seals

This involves partnering with companies or brands relevant to your podcast's niche. You can integrate sponsored messages or segments into your episodes, and in return, you receive payment from the sponsor.

If your podcast concerns health and fitness, you can partner with a fitness equipment company to promote their products or include sponsored segments about their latest fitness gear in your episodes.

Become a Sponsor for Another Podcast

If you have a successful podcast, you can invest in sponsoring another podcast with a similar target audience, increasing your visibility and potentially attracting new listeners.

Sponsorship deals may allow you to get a cut of total revenue by making you a "partner." Additionally, similar-minded podcasters pool their resources to create "networks" that cast a much wider net. For instance, sports podcasters focusing on basketball work together to cover the sport but concentrate on different teams since they live in other cities or countries.

Use Affiliate Marketing

Promote products or services through unique affiliate links in your episodes. When listeners make purchases through those links, you earn a commission.

Suppose your podcast covers technology. You can become an affiliate for an online electronics retailer and promote their products in your episodes, including unique affiliate links. When listeners use those links to make purchases, you earn a commission on each sale.

Sell Show Merchandise

Create branded merchandise such as t-shirts, mugs, or stickers featuring your podcast's logo or catchphrases. Sell these items to your loyal listeners.

If you host a comedy podcast with catchphrases and inside jokes, design humorous t-shirts or coffee mugs featuring these catchphrases and offer them for sale on your website or through an e-commerce platform.

Sell repurposed content

Package and sell transcripts, bonus episodes, or exclusive content that didn't make it into your regular episodes.

Additionally, some podcasters make unreleased outtakes or unaired episodes available to subscribers only. This exclusive content, while not part of the usual show's issue, mainly contains a unique behind-the-scenes look, enabling listeners to get a peek at content not available to free subscribers or the general public.

Offer Premium Paid Content

Establish a subscription-based model where listeners pay a monthly fee to access premium episodes or content.

If you run a podcast about personal finance, you could create a subscription-based model where subscribers get access to advanced financial planning webinars or one-on-one financial coaching sessions.

Create Tiered Premium Content

Offer different subscription tiers with varying levels of benefits, such as early access, exclusive Q&A sessions, or behind-the-scenes content.

This approach can be tricky. So, it's essential to structure your content. Hence, subscribers know paying extra for premium content is worth the added expense. For example, a health and wellness podcast makes free presentations available to the general public in which topics do not enter into great depth. The next premium tier could focus on content exclusive to specific topics. In contrast, the highest premium tier could focus on live Q&A sessions directly addressing audience concerns.

Accept Donations

Allow your audience to support your podcast voluntarily through platforms like Patreon or Ko-fi, offering special perks or recognition to donors.

Set up a Patreon account for your podcast and offer donors perks like shout-outs in episodes, exclusive content, or early access to episodes. Encourage listeners to contribute voluntarily to support your show.

Join an Advertising Network

Partner with podcast advertising networks that connect you with advertisers and handle ad placement in your episodes.

These advertising networks generally pay a fixed amount per ad seen or heard by audiences. While individual payments, generally pennies per view or listen, may seem irrelevant, it can become a solid investment when multiplied over thousands or millions of views and listens. Large e-commerce platforms like Amazon offer such opportunities, while some individual brands may have similar options.

Syndicate Your Show to YouTube

Convert your podcast episodes into video format and upload them to YouTube. Monetize through YouTube's ad revenue, sponsorships, or affiliate marketing.

Convert your podcast episodes into video format and upload them to YouTube. You can monetize by enabling YouTube ads on your videos, earning revenue based on ad views and clicks.

Public Speaking

Leverage your podcast's expertise and reputation to secure speaking engagements at conferences or events, where you can earn speaking fees.

If you plan on public speaking, let your audience know you're open to them. To do so, include an email address available for business inquiries. Having a business manager field these requests can help ease the burden as you focus on creating high-quality content.

Sell Access to an E-course

Develop an educational course related to your podcast's niche and offer it for a fee to your listeners. These courses should provide something that your podcast does not. For instance, an e-course can extensively examine a particular topic.

E-courses are very popular among a wide range of topics. However, you must ensure your e-course covers topics that would be too extensive or complex for your podcast. This situation compels listeners to sign up for your course instead of only listening to your show.

Sell Mastermind Slots

Create a mastermind group or coaching program where listeners can participate for a fee to gain access to your knowledge and network.

Group and individual coaching sessions are very effective at monetizing your podcast. For example, podcasters in personal development areas can make themselves available to listeners via Zoom. However, you must be mindful of your time, especially if you get flooded with requests. Be sure to let your audience know when you're available and for how long. People won't mind being on a waiting list if they know what to expect.

Host an Event

Organize live events, workshops, or webinars related to your podcast's content and sell tickets or access passes.

Organize a virtual business summit featuring guest speakers and workshops if your podcast focuses on entrepreneurship. Sell tickets or access passes for attendees to gain insights and networking opportunities.

Sell Affiliate Products

Promote and earn commissions from products or services that align with your podcast's theme.

Suppose you host a podcast about cooking. Recommend kitchen appliances and utensils you personally use and love, and provide affiliate links to online stores where listeners can purchase them.

Create Your Community

Establish a private online community or forum for your podcast listeners, offering exclusive content and charging a membership fee.

Establish a private Facebook group for your podcast's community, where members can engage in discussions, access bonus content, and participate in live Q&A sessions. Charge a monthly membership fee for access to this exclusive community.

Remember that the success of these monetization methods often depends on your podcast's niche, audience size, and engagement level. Choosing methods that align with your content and provide value to your listeners while being transparent about your monetization efforts to maintain trust with your audience is essential.

INTELLECTUAL PROPERTY AND COPYRIGHTS

[3]Like any content creator, podcasters should be aware of various legal considerations and intellectual property rights to ensure they operate within the law's bounds and avoid potential legal issues. Here's a detailed explanation of each of these legal aspects:[4]

Intellectual Property Law

Intellectual property (IP) law encompasses various legal protections for creative works and innovations. For podcasters, the critical elements of IP law to understand are trademarks, patents, and copyrights. Trademarks protect brand names and logos, patents protect inventions, and copyrights protect original creative works. In the context of podcasting, copyrights are particularly important.

Copyright Law

Copyright law grants the creator of an original work exclusive rights to its use and distribution. For podcasters, the content you create, such as your episodes, scripts, music, and artwork, is automatically protected by copyright as soon as it's made and fixed in a tangible medium (e.g., recorded). You have the exclusive right to reproduce, distribute, and display your work.

Podcasters should also be aware of the rights of others' copyrighted works. Using copyrighted material, like music, without permission can lead to copyright infringement claims. To avoid this, you should obtain the necessary licenses or use royalty-free music and ensure your content falls under fair use (explained below) when using copyrighted material.

Privacy Law

Privacy laws, including data protection regulations like GDPR in Europe and various state privacy laws in the United States, protect the personal information of individuals. When podcasting, you should be cautious about collecting, storing, and sharing personal data. If you interview guests or collect listener data, you must obtain consent and handle the data in compliance with relevant privacy laws.

Public Domain

The public domain refers to creative works no longer protected by copyright and are free for anyone to use without permission. Works can enter the public domain through various means, such as when the copyright expires (usually after 70 years following the creator's death), the creator explicitly waives their copyright, or the work is never eligible for copyright protection.

Podcasters need to ensure that any content they use from the public domain is genuinely in the public domain, as mistakenly using copyrighted material can lead to legal issues.

Fair Use Defense

Fair use is a legal doctrine in copyright law that allows limited use of copyrighted material without permission for purposes such as criticism, commentary, news reporting, education, or parody. Whether a particular use qualifies as fair use depends on several factors, including the purpose and character of the use, the nature of the copyrighted work, the amount used, and the effect on the market value of the copyrighted work.

While fair use provides some flexibility for podcasters to use copyrighted material, it is a complex and often subjective area of law. It's crucial to consult with legal counsel or understand fair use principles thoroughly before relying on it as a defense.

Keep in mind that failing to follow these guidelines can lead to significant legal consequences such as lawsuits. Additionally, platforms such as YouTube can demonetize and take down channels due to "copyright strikes." Similarly, directories like Spotify adhere strictly to copyright laws, banning shows known to break these laws. Therefore, ensuring your show follows guidelines and applicable laws is crucial to your show's success.

Other Legal Compliance Considerations for Podcasters

The following legal considerations are crucial for podcasters to follow in addition to copyright and intellectual property guidelines:

Defamation Laws

Defamation laws protect individuals and entities from false statements that harm their reputations. In podcasting, defamation can occur if you make false statements about someone, damaging their character or reputation. You do so negligently or with actual malice (knowingly false statements or reckless disregard for the truth).

To comply with defamation laws, podcasters should:

- Ensure that statements made on their podcast are factual or based on reliable sources.
- Avoid making false or unsubstantiated claims about individuals, companies, or products.
- Exercise caution when discussing sensitive or controversial topics, especially involving individuals.

Defamation laws vary by jurisdiction, so it's essential to be familiar with the laws in your region and consult legal counsel if you have concerns about potentially defamatory content.

Suppression Orders

Suppression orders, also known as gag orders or publication bans, are legal orders issued by a court to restrict the publication of specific information. These orders are typically used in cases involving sensitive matters like ongoing investigations, national security, or the protection of vulnerable individuals.

Podcasters should be aware of suppression orders in their jurisdiction and comply with them if they receive a court-issued order. Failing to adhere to suppression orders can result in legal consequences.

Guest Release Forms

Guest release forms are legal documents signed by podcast guests, granting permission for their likeness, voice, and content to be used in the podcast episode. These forms are essential to protect podcasters from potential legal disputes and ensure they have the right to publish the content.

A guest release form typically includes the following:

- Permission to record and use the guest's voice and image.
- Agreement to the terms of the interview, including any sensitive topics.

- Waiver of any claims or rights to compensation for participating in the podcast.

Podcasters should have a clear and well-drafted guest release form signed by every guest to avoid legal issues related to consent and rights to the content.

Disclaimers

Disclaimers are statements or notices included in podcast episodes or show descriptions to clarify certain aspects of the content, such as opinions, sponsorship, or potential legal issues. While disclaimers may not provide complete legal protection, they can help manage expectations and reduce the risk of misunderstandings.

Common types of disclaimers for podcasters include:

- **Opinions.** Stating that the views expressed in the podcast are the opinions of the hosts or guests.
- **Sponsorship.** Disclosing any paid partnerships or sponsored content.
- **Legal advice.** Reminding listeners that the podcast is not a substitute for legal, medical, or professional advice.
- **Explicit content.** Alerting listeners to explicit language or mature themes.

Disclaimers should be clear, concise, and tailored to the podcast's content and potential legal concerns.

Overall, you should be diligent in understanding and complying with these legal considerations, both as a courtesy to your audience and guests and to avoid needless legal drama. Adherence to legal guidelines shows your audience you take ethics seriously.[5]

CONSEQUENCES OF LEGAL VIOLATIONS

Legal violations can have serious consequences, ranging from lawsuits to criminal charges. The following situations exemplify the ramifications of these issues:[6]

Defamation and Libel

In a podcast episode, a host made false and damaging statements about a local business owner, alleging criminal activities without any evidence to support the claims. The episode received a substantial number of downloads and gained attention on social media.

Legal Consequences: The business owner sued the podcaster for defamation and libel. The court found that the statements were false and made with reckless disregard for the truth. As a result, the podcaster was ordered to pay significant damages to the business owner and had to issue a public apology on their podcast, acknowledging the false statements.

Copyright Infringement

A podcaster used copyrighted music without obtaining the necessary licenses or permissions in several of their episodes. The music was used as background music for various segments and intros.

Legal Consequences: The copyright holders of the music tracks issued takedown notices, and the podcaster's episodes containing the copyrighted music were removed from podcasting platforms. Additionally, the podcaster received cease-and-desist letters and faced potential lawsuits for copyright infringement, which could result in substantial fines and damages.

Privacy Violation

In an attempt to boost engagement, a podcaster revealed sensitive personal information about a listener who had written in with a question, including their full name, location, and a description of their situation.

Legal Consequences: The listener filed a privacy lawsuit against the podcaster for violating their personal information without consent. The podcaster faced legal action, potential damages for the privacy violation, and damage to their podcast's reputation due to the breach of trust with their audience.

These examples illustrate the serious legal consequences podcasters can face when they engage in activities such as defamation, copyright infringement, or privacy violations. Legal actions can result in financial penalties, damage to one's reputation, and even the shutdown of the podcast. To avoid these consequences, podcasters should always strive to create content that is legally compliant, respectful of others' rights, and considerate of privacy and intellectual property laws.

BRINGING IT ALL TOGETHER

Monetizing your podcast is undeniably a key metric for measuring its overall success and sustainability. It signifies that your content resonates with an audience and can generate income, validating the time and effort invested. However, this path to financial gain must be tread carefully, with meticulous attention to legal compliance.

While monetization offers exciting opportunities, it also comes with a web of legal considerations. Podcasters must navigate a complex legal landscape from copyright and defamation concerns to privacy issues and advertising regulations. Failure to do so can lead to costly legal disputes, damage to reputation, and even the shutdown of a thriving podcast.

To ensure long-term prosperity and mitigate legal risks, podcasters should prioritize compliance with intellectual property laws, transparency in advertising and sponsorship deals, and the respectful treatment of guest consent and privacy. By striking a balance between monetization and legal diligence, podcasters can reap the rewards of their hard work while safeguarding their content's integrity and future success.

CONCLUSION

*The real value is the conversations and relation-
ships built from your podcast.*

— YIFAT COHEN

As we draw to the end of this book, we have seen how
podcasting has emerged as one of the most rewarding
endeavors one can embark upon in this digital age. The
journey from concept to creation, and eventually
connecting with a global audience, offers an unparalleled
sense of fulfillment. Throughout this book, we have
delved into the multifaceted world of podcasting,
revealing its vast potential for anyone willing to seize the
microphone.

Podcasting is not just about broadcasting your voice; it's
about building connections, sharing stories, and fostering

a sense of community. As we've explored the tips, strategies, and guidelines within these pages, it's evident that podcasting is accessible to all, regardless of your background or expertise. Your unique perspective and passion are the only prerequisites for starting your podcasting journey.

While the prospect of launching a podcast may appear daunting, remember that every great podcaster was once a beginner. The knowledge gained from this book will serve as a springboard, propelling you into the dynamic world of podcasting with confidence. Whether you aspire to inform, inspire, educate, entertain, or engage in meaningful conversations, podcasting provides a platform for your voice to be heard.

Podcasting offers the opportunity to transcend boundaries, share your insights, and connect with like-minded individuals around the globe. It's a medium that rewards the creators and listeners who find value, solace, or joy in your content. As you embark on your podcasting journey, remember that your message matters, and your voice deserves to be amplified. So, embrace the adventure, harness the power of podcasting, and let your unique story resonate with the world.

But what if you've already tried podcasting and have failed?

Perhaps you started enthusiastically and poured your heart and soul into it, only to face discouraging download numbers or waning interest. It's essential to remember that setbacks are a natural part of any creative journey. With this book, you have a beacon of hope to revive your podcasting dreams.

Throughout these pages, we discussed a treasure trove of strategies, insights, and actionable tips to breathe new life into your struggling podcast. We understand that it can be disheartening to see your efforts go seemingly unnoticed. Still, the worst thing you can do is give up on your endeavors. The world deserves to hear your message, and there are listeners out there who need your unique perspective and insights.

This book equips you with the knowledge and tools to analyze what might have gone wrong, refine your content, and re-engage your audience. From rebranding to redefining your niche, we'll guide you through the steps to rejuvenate your podcast and recapture the enthusiasm that drove you to start it in the first place.

Reviving a seemingly failed podcast is not just about boosting download numbers; it's about reconnecting with your passion and purpose. The journey of podcasting is filled with ups and downs. Still, it's the resilience and commitment that set successful podcasters apart. Your voice matters, and the world is waiting to hear your message. As you absorb the insights from this book, remember that every setback is an opportunity for a comeback. Embrace the challenge, harness the wisdom within these pages, and let your podcast shine once more.

But what if you're still not convinced podcasting can work for you?

Let me share a story to fuel your motivation:

Everybody has heard of Joe Rogan. It's hard to imagine that there's someone who hasn't heard of him. In fact, when you think of successful podcasts, Rogan is the first name that comes to mind. Rogan's meteoric rise to becoming the world's most famous podcaster can be attributed to a combination of factors that have set his show apart and made it an unparalleled success.

But why has his show been so successful?

Consider these elements:

- **Authentic Conversations.** Joe Rogan's podcast, "The Joe Rogan Experience," is known for its authenticity. He creates a relaxed, open atmosphere where guests feel comfortable sharing their thoughts and experiences. This authenticity fosters genuine, unscripted conversations that resonate with listeners.
- **Diverse Range of Guests.** Rogan's show features a diverse lineup of guests, ranging from celebrities and comedians to scientists, authors, and experts in various fields. This diversity

ensures a broad appeal, drawing in listeners with varied interests.

- **Long-Form Content.** Unlike many podcasts with strict time constraints, Rogan's episodes often stretch for several hours. This long-form format allows for in-depth exploration of topics. It encourages a deeper connection between the host, guests, and the audience.
- **Fearless Exploration of Controversial Topics.** Rogan is not afraid to tackle controversial or taboo subjects. His willingness to engage in open and honest discussions about sensitive issues has earned him a reputation for fearlessness and intellectual curiosity.
- **Relatability.** Despite his fame, Joe Rogan maintains a relatable persona. Listeners feel like they're having a conversation with a friend, which enhances the show's appeal and fosters a sense of connection.
- **High Production Quality.** Rogan's podcast benefits from top-notch production quality, with excellent sound, video, and a professional studio setup. This commitment to quality enhances the overall listening experience.
- **Strong Social Media Presence.** Joe Rogan leverages social media platforms to promote his podcast, connect with his audience, and share clips and highlights from his show, further expanding his reach.

- **Consistency.** Rogan has consistently produced new episodes regularly, which builds his listeners' anticipation and loyalty.

While Rogan embodies a number of highly successful traits needed to succeed in the podcasting world, arguably, his biggest claim to fame is his work ethic and consistency. Rogan's dedication to his craft is legendary. He's known to put in the time and effort to make his show as successful as it can be. He reads, studies various subjects, and learns about his guests.

Plus, Rogan listens to his audience. He brings people in that resonate with his audiences. But they are not the same type of guests. The guests that come on his show are voices people want to hear. They are often people involved in trending topics. But they are also people with differing points of view. As a result, the richness of opinions makes Rogan's podcast the most popular of all time.

So, take a page out of Rogan's book. Don't be afraid to put yourself out there. Allow yourself to make the most of the wonderful technology at your disposal. Unlike any other time in human history, we have the means to broadcast a message to the entire world.

It takes time and patience.

Joe Rogan didn't become an instant star. He needed to put the work in to get to where he is today. But that didn't stop him. His passion fueled his drive to become the best at what he does. But do you know what made his success possible? He didn't think of his show in terms of a commercial enterprise. He viewed his show as an opportunity to express his passion for the things he truly believes in.

You have the same opportunity, too.

Thank you very much for taking the time to read this book. The end of this discussion is by no means the end of the road. If anything, it is merely the beginning of a wonderful journey that will take you places you may not have thought about before.

Now is the time to seize the day and make the most of this unique opportunity to get your message out to the world. Please take the guidelines we have provided in this book. They will surely become your guiding beacon throughout your journey.

Thank you once again for your time and attention. We'll be looking forward to hearing you on your show!

NOTES

1. UNDERSTANDING PODCASTING

1. McGarr, Oliver. "A review of podcasting in higher education: Its influence on the traditional lecture." *Australasian journal of educational technology* 25, no. 3 (2009).
2. *Ibid.*
3. "How Podcasting Has Grown Over The Years." MartechSeries, November 30, 2022. https://martechseries.com/mts-insights/staff-writers/how-podcasting-has-grown-over-the-years/.
4. *Ibid.*
5. *Ibid.*
6. *Ibid.*
7. "How Podcasting Has Grown Over The Years." MartechSeries, November 30, 2022. https://martechseries.com/mts-insights/staff-writers/how-podcasting-has-grown-over-the-years/.
8. Flanagan, Brian, and Brendan Calandra. "Podcasting in the classroom." *Learning & Leading with Technology* 33, no. 3 (2005): 20-23.
9. "How Podcasting Has Grown Over The Years." MartechSeries, November 30, 2022. https://martechseries.com/mts-insights/staff-writers/how-podcasting-has-grown-over-the-years/.
10. Campbell, Cara. "THE STRUGGLE IS REAL: THE IMPOSTER SYNDROME." PhD diss., California State University, Chico, 2021.
11. Wilding, Melody J. "5 Types of Imposter Syndrome and How to Stop Them." The Muse, May 10, 2017. https://www.themuse.com/advice/5-different-types-of-imposter-syndrome-and-5-ways-to-battle-each-one.
12. *Ibid.*
13. Pogored. "What's Imposter Syndrome and How to Overcome It." Cleveland Clinic, April 4, 2022. https://health.clevelandclinic.org/a-psychologist-explains-how-to-deal-with-imposter-syndrome/.

2. DEVELOPING A COMPELLING PODCAST CONCEPT

1. Haahr, Tae. "How to Find Your Podcast Niche." The Podcast Host, June 28, 2021. https://www.thepodcasthost.com/planning/podcast-niche/.
2. "How to Find the Perfect Podcast Niche for Your Show." Podcast.co. Accessed August 5, 2023. https://blog.podcast.co/reach/find-podcast-niche.
3. McNamee, Lyn. "How to Find a Winning Podcast Niche (Examples & Ideas)." The Rephonic Blog, September 7, 2022. https://rephonic.com/blog/finding-a-podcast-niche.
4. Molenaar, Koba. "40 Podcast Topic Ideas to Grow Your Podcast in 2023." Influencer Marketing Hub, December 7, 2022. https://influencermarketinghub.com/podcast-topic-ideas/.
5. "100+ Creative Podcast Topics Ideas in 2023." Riverside. Accessed August 5, 2023. https://riverside.fm/blog/podcast-ideas-to-try.
6. "How to Discover and Reach Your Podcast's Target Audience." RSS. Accessed August 5, 2023. https://podcasters.spotify.com/resources/learn/grow/podcast-target-audience.
7. *Ibid.*
8. DOĞAN, Şeyhmus. "CURRENT MARKETING METHODS: ANALYSIS OF DIMENSIONS AND APPLICATION STRATEGIES OF NICHE MARKETING CONCEPT." *New Horizons in Social, Human and Administrative Sciences* 479.
9. Pitsillis, Giorgos. "Organic vs. paid social media marketing." (2023).
10. *Ibid.*
11. Clifton, Rita. *Brands and branding.* Vol. 43. John Wiley & Sons, 2009.
12. *Ibid.*
13. Team, Podcastle. "Picking a Good Podcast Name - Top Tips!" Podcastle Blog, August 7, 2023. https://podcastle.ai/blog/how-to-create-a-relevant-podcast-name-for-your-audience.

3. ESSENTIAL PODCASTING EQUIPMENT AND SOFTWARE

1. Wolff, Terris B. "Podcasting made simple." In *Proceedings of the 34th annual ACM SIGUCCS fall conference: expanding the boundaries*, pp. 413-418. 2006.
2. Hurst, Emily J. "Getting started with podcasting." *Journal of Hospital Librarianship* 19, no. 3 (2019): 277-283.
3. *Ibid.*
4. "Best Podcast Equipment: Must-Haves for Any Budget Setup (2023)." Riverside. Accessed August 7, 2023. https://riverside.fm/blog/podcast-equipment.
5. Hurst, *ibid.*
6. Best Podcast Equipment, *ibid.*
7. Corfield, Chris. "Best Podcast Mixers 2023: The Central Hub of Your Podcasting Setup." MusicRadar, May 9, 2022. https://www.musicradar.com/news/best-podcast-mixers.
8. Best Podcast Equipment, *ibid.*
9. Light, Colbor. "Podcast Light: Three Basics to Know for Better Podcasts." COLBOR, July 23, 2023. https://www.colborlight.com/blogs/articles/podcast-light-basics.
10. Best Podcast Equipment, *ibid.*
11. "The 8 Best Pop Filters for Podcasters in 2023 (Every Budget)." Riverside. Accessed August 7, 2023. https://riverside.fm/blog/best-pop-filter.
12. Best Podcast Equipment, *ibid.*
13. *Ibid.*
14. "Best Digital Audio Recorders for Podcasting in 2022." Keynote Content, January 11, 2022. https://www.keynotecontent.com/best-digital-audio-recorders-podcasting-2022/.
15. Best Podcast Equipment, *ibid.*
16. Watson, Zach. "All the Podcast Software You Need to Create A Killer Show." Soundstripe Royalty Free Music, April 11, 2023. https://www.soundstripe.com/blogs/podcast-software.
17. Podcast equipment: Everything you need to build your studio. Accessed August 7, 2023. https://www.adorama.com/alc/podcast-equipment-everything-you-need-to-build-your-studio/.

18. "11 Best Audio Recording Software for All Budgets & PCs in 2023." Riverside. Accessed August 7, 2023. https://riverside.fm/blog/audio-recording-software.

19. Team, Podcastle. "How to Create the Ultimate Podcast Studio." Podcastle Blog, October 17, 2023. https://podcastle.ai/blog/how-to-create-the-ultimate-podcast-studio/.

4. RECORDING AND EDITING YOUR PODCAST

1. 15 microphone techniques you should be using | podcast.co. Accessed August 7, 2023. https://blog.podcast.co/create/microphone-techniques.

2. Jeff, and Roy Patterson. "Podcast Mic Techniques for New Hosts (11 Important Tips)." Castos, May 8, 2023. https://castos.com/mic-techniques/.

3. Sastry, Shashidhar, Geekupilsa, and Shannon Peel. "Podcast Recording Tips for Polished, Professional Episodes." Castos, May 23, 2023. https://castos.com/podcast-recording-tips/.

4. "Podcast Editing: How to Do It in 10 Steps (Complete Tutorial)." Riverside. Accessed August 11, 2023. https://riverside.fm/blog/podcast-editing.

5. Street, Lower. "How to Edit a Podcast: Audio Levels, Content, Tips, Trailers & More: Lower Street." Podcast Production Services | Lower Street. Accessed August 11, 2023. https://lowerstreet.co/how-to/edit-a-podcast.

6. "Podcast Editing: How to Do It in 10 Steps (Complete Tutorial)." Riverside. Accessed August 11, 2023. https://riverside.fm/blog/podcast-editing.

5. STRUCTURING YOUR PODCAST EPISODES

1. Golobe, Mike, Erin Kratz, Janell, Chris Land, Henry I Padrón-Morales, and The Dysfunktionals. "How to Plan Podcast Episode

Structure in 10 Easy Steps." Improve Podcast, March 23, 2022. https://improvepodcast.com/9-key-tips-on-how-to-plan-podcast-episodes/.

2. *Ibid.*

3. Buzzsprout. "How to Structure Your Podcast in 5 Steps." Buzzsprout. Accessed August 11, 2023. https://www.buzzsprout.com/blog/podcast-structure.

4. Street, Lower. "How to Structure a Podcast to Keep Listeners Engaged: Lower Street." Podcast Production Services | Lower Street. Accessed August 11, 2023. https://lowerstreet.co/how-to/structure-podcast.

5. Rob ScheerbarthRob Scheerbarth is the owner of Podcast Rocket. "How to Create an Engaging Podcast Intro (+examples!)." Podcast Rocket, February 4, 2023. https://podcastrocket.net/how-to-create-a-podcast-intro/.

6. Ivanescu, Yvonne. "How to Create the Perfect Podcast Intro & Outro." Spreaker Blog, March 1, 2022. https://blog.spreaker.com/how-to-create-the-perfect-podcast-intro-outro/.

7. Slack, Alexandra, Macks, Latisha Lewis, Wunmi Thegreat, Renata Franklin, Julio A Barrios Lemole, Rob Weddle, and Sam Chlebowski. "Engaging Podcast Content: 13 Tips to Create Better Content." Castos, July 4, 2023. https://castos.com/engaging-podcast-content/.

8. Land, Chris. "Storytelling in Podcasting: 10 Steps to Boost Engagement by 300%." Improve Podcast, November 25, 2021. https://improvepodcast.com/storytelling-in-podcasting/.

9. Ali. "How to Write a Good Podcast Conclusion?: Ausha Blog." Ausha, October 24, 2023. https://www.ausha.co/blog/write-podcast-conclusion/.

10. Heerden, Carli van. "8 Best Practices for Podcast Interviews: We Edit Podcasts." We Edit Podcasts | A Podcast Production Agency, August 24, 2023. https://weeditpodcasts.com/8-best-practices-for-podcast-interviews/.

6. PODCAST HOSTING AND DISTRIBUTION

1. "Podcast Host Directory Website...the Differences." Podcast Pursuit, June 23, 2023. https://www.podcastpursuit.com/differ ence-podcast-host-directory-website/#Podcast-Host-Directory-Website-An-Overview-of-the-Differences.
2. "What's the Difference between a Podcast Host and Directory?" Podcast Services Australia, September 23, 2021. https://podcastser vices.com.au/whats-the-difference-between-a-podcast-host-and-directory/.
3. Podcast Host Directory Website, *ibid.*
4. "What to Consider When Choosing a Podcast Hosting Platform." Riverside. Accessed August 15, 2023. https://riverside.fm/blog/choosing-a-podcast-hosting-platform.
5. *Ibid.*
6. *Ibid.*
7. Media, Dear. "How to Choose a Podcast Hosting Platform - Dear Media." Dear Media - New Way to Podcast, August 29, 2022. https://dearmedia.com/how-to-choose-a-podcast-hosting-plat form/.
8. "Podcast Host Directory Website...the Differences." Podcast Pursuit, June 23, 2023. https://www.podcastpursuit.com/differ ence-podcast-host-directory-website/.
9. "How to Publish a Podcast (Beginner's Guide 2023)." Riverside. Accessed August 15, 2023. https://riverside.fm/blog/how-to-publish-a-podcast.
10. *Ibid.*
11. *Ibid.*
12. Lee, Kevan. "Podcasting for Beginners: The Complete Guide to Getting Started." Buffer Library, February 27, 2023. https://buffer. com/library/podcasting-for-beginners/.
13. Russell, Izabela. "How to Optimize Your Podcast Metadata - Detailed Explanation." Music Radio Creative, September 6, 2023. https://producer.musicradiocreative.com/how-to-optimize-your-podcast-metadata/.
14. *Ibid.*

15. "How to Optimize Your Podcast's Metadata: Ausha Blog." Ausha, September 26, 2023. https://www.ausha.co/blog/optimize-podcast-metadata/.

7. EFFECTIVE MARKETING AND PROMOTION STRATEGIES

1. OToole, Terry. "Key Strategies to Build a Podcast Brand." Tailor Brands, July 27, 2022. https://www.tailorbrands.com/blog/build-a-podcast-brand.
2. "Council Post: 16 pro Tips for Creating a Successful Branded Podcast." Forbes, November 21, 2022. https://www.forbes.com/sites/forbesagencycouncil/2022/11/18/16-pro-tips-for-creating-a-successful-branded-podcast/?sh=5d75ed6b7dfc.
3. Brinson, Nancy H., and Laura L. Lemon. "Investigating the effects of host trust, credibility, and authenticity in podcast advertising." *Journal of Marketing Communications* 29, no. 6 (2023): 558-576.
4. Dubey, Shashank. "Nurture Your Podcast Audience with Online Community." Wbcom Designs, January 12, 2023. https://wbcomdesigns.com/nurture-your-podcast-audience/.
5. "Listener Engagement Advice from 5 Podcast Experts." Podcast.co. Accessed August 25, 2023. https://blog.podcast.co/reach/podcast-listener-engagement#scroll.
6. 12 Best-Practices to Make Podcast Cover Art That Works." Riverside. Accessed August 25, 2023. https://riverside.fm/blog/podcast-cover-art.
7. "How to Design Unique & Distinctive Podcast Cover Art." Podcast.co. Accessed August 25, 2023. https://blog.podcast.co/create/podcast-cover-art-logo.
8. Team, Podcastle. "How to Write a Podcast Description: Examples to Help You Write Your Own." Podcastle Blog, October 24, 2023. https://podcastle.ai/blog/how-to-write-a-podcast-description/.
9. "How to Write Amazing Podcast Descriptions (with Examples)." Riverside. Accessed August 25, 2023. https://riverside.fm/blog/podcast-descriptions.
10. Podcast, *Ibid.*

11. Riverside, *ibid.*
12. "10 Effective Strategies for Podcast Social Media Promotion." Castmagic. Accessed August 26, 2023. https://www.castmagic.io/post/10-effective-strategies-for-podcast-social-media-promotion.
13. Townsend, Stewart. "Podcast Marketing as a Guest: Leveraging Social Media -." Podcast Hawk [Beta] - Find 1000s of podcast interview opportunities in seconds, May 10, 2023. https://podcasthawk.com/podcast-marketing-as-a-guest-leveraging-social-media/.
14. "8 Effective Podcast Collaboration Strategies to Grow Your Show." Castos, May 7, 2023. https://castos.com/podcast-collaboration/.
15. "How Podcast Collaborations Can Grow Your Reach." RSS. Accessed August 72, 2023. https://www.bcast.fm/blog/podcast-collaborations.
16. McDowell, Sarah. "Podcast Seo: 10 Tips from Our SEO Podcasting Expert: Captivate." Captivate Unlimited Podcast Hosting & Analytics, August 10, 2023. https://www.captivate.fm/podcast-growth/marketing/10-insider-podcast-seo-tips-to-opti mize-your-podcast.
17. "Unlock the Power of Podcast SEO: 5 Strategies for Content Creation Success." Truth Work Media, March 29, 2023. https://truthworkmedia.com/podcast-seo/.

8. MONETIZING YOUR PODCAST

1. Louis, Belkacem, Sandra Urquhart, ngan LT, Riss, Amel Snv, ferahtia.Fs, Tita, Vincent, and Podcast VirtualAssistant. "How Do Podcasters Make Money?: 20 Strategies to Monetize a Podcast." Castos, July 3, 2023. https://castos.com/monetize-a-podcast/.
2. Cook, Jodie. "How to Monetize Your Podcast: 6 Proven Methods." Forbes, October 12, 2022. https://www.forbes.com/sites/jodiecook/2022/10/10/how-to-monetize-your-podcast-6-proven-methods/?sh=67be0dd16e4c.
3. "Top Legal Considerations for Podcasters." The Browne Firm, September 21, 2021. https://www.thebrownefirm.com/top-legal-considerations-for-podcasters.

4. Team, Podcastle. "5 Podcasting Copyright Laws." Podcastle Blog, September 12, 2022. https://podcastle.ai/blog/podcasting-copyright-laws/.

5. "7 Laws Every Podcast Creator Needs to Know." The Simplecast Blog, June 22, 2023. https://blog.simplecast.com/7-podcasting-laws.

6. Weitmann, Deserae. "Podcasting Legal Issues: Avoiding Infringement with a Podcast." CreativeLive Blog, April 12, 2016. https://www.creativelive.com/blog/podcast-legal-issues/.